# Selected Titles in This Series

D1805568

732 **Jan O. Kleppe, Juan C. Migliore, Rosa Miró-Roig, Uwe Nagel, and Chris Peterson,** Gorenstein liaison, complete intersection liaison invariants and unobstructedness, 2001

731 **Jesús Bastero, Mario Milman, and Francisco J. Ruiz,** On the connection between weighted norm inequalities, commutators and real interpolation, 2001

730 **Suhyoung Choi,** The decomposition and classification of radiant affine 3-manifolds, 2001

729 **Michael Grosser, Eva Farkas, Michael Kunzinger, and Roland Steinbauer,** On the foundations of nonlinear generalized functions I and II, 2001

728 **Laura Smithies,** Equivariant analytic localization of group representations, 2001

727 **Anthony D. Blaom,** A geometric setting for Hamiltonian perturbation theory, 2001

726 **Victor L. Shapiro,** Singular quasilinearity and higher eigenvalues, 2001

725 **Jean-Pierre Rosay and Edgar Lee Stout,** Strong boundary values, analytic functionals, and nonlinear Paley-Wiener theory, 2001

724 **Lisa Carbone,** Non-uniform lattices on uniform trees, 2001

723 **Deborah M. King and John B. Strantzen,** Maximum entropy of cycles of even period, 2001

722 **Hernán Cendra, Jerrold E. Marsden, and Tudor S. Ratiu,** Lagrangian reduction by stages, 2001

721 **Ingrid C. Bauer,** Surfaces with $K^2 = 7$ and $p_g = 4$, 2001

720 **Palle E. T. Jorgensen,** Ruelle operators: Functions which are harmonic with respect to a transfer operator, 2001

719 **Steve Hofmann and John L. Lewis,** The Dirichlet problem for parabolic operators with singular drift terms, 2001

718 **Bernhard Lani-Wayda,** Wandering solutions of delay equations with sine-like feedback, 2001

717 **Ron Brown,** Frobenius groups and classical maximal orders, 2001

716 **John H. Palmieri,** Stable homotopy over the Steenrod algebra, 2001

715 **W. N. Everitt and L. Markus,** Multi-interval linear ordinary boundary value problems and complex symplectic algebra, 2001

714 **Earl Berkson, Jean Bourgain, and Aleksander Pełczynski,** Canonical Sobolev projections of weak type $(1, 1)$, 2001

713 **Dorina Mitrea, Marius Mitrea, and Michael Taylor,** Layer potentials, the Hodge Laplacian, and global boundary problems in nonsmooth Riemannian manifolds, 2001

712 **Raúl E. Curto and Woo Young Lee,** Joint hyponormality of Toeplitz pairs, 2001

711 **V. G. Kac, C. Martinez, and E. Zelmanov,** Graded simple Jordan superalgebras of growth one, 2001

710 **Brian Marcus and Selim Tuncel,** Resolving Markov chains onto Bernoulli shifts via positive polynomials, 2001

709 **B. V. Rajarama Bhat,** Cocylces of CCR flows, 2001

708 **William M. Kantor and Ákos Seress,** Black box classical groups, 2001

707 **Henning Krause,** The spectrum of a module category, 2001

706 **Jonathan Brundan, Richard Dipper, and Alexander Kleshchev,** Quantum Linear groups and representations of $GL_n(\mathbb{F}_q)$, 2001

705 **I. Moerdijk and J. J. C. Vermeulen,** Proper maps of toposes, 2000

704 **Jeff Hooper, Victor Snaith, and Min van Tran,** The second Chinburg conjecture for quaternion fields, 2000

703 **Erik Guentner, Nigel Higson, and Jody Trout,** Equivariant $E$-theory for $C^*$-algebras, 2000

*(Continued in the back of this publication)*

# Gorenstein Liaison, Complete Intersection Liaison Invariants and Unobstructedness

# MEMOIRS
of the
American Mathematical Society

Number 732


# Gorenstein Liaison, Complete Intersection Liaison Invariants and Unobstructedness


Jan O. Kleppe
Juan C. Migliore
Rosa Miró-Roig
Uwe Nagel
Chris Peterson



November 2001 • Volume 154 • Number 732 (third of 5 numbers) • ISSN 0065-9266

**American Mathematical Society**
Providence, Rhode Island

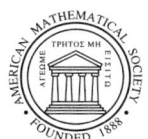

2000 *Mathematics Subject Classification.*
Primary 14M12, 14C05, 14H10, 14J15; Secondary 14N05.

---

**Library of Congress Cataloging-in-Publication Data**

Gorenstein liaison, complete intersection liaison invariants, and unobstructedness / Jan O. Kleppe...[et al.].

p. cm. — (Memoirs of the American Mathematical Society, ISSN 0065-9266 ; no. 732)

"Volume 154, number 732 (third of 5 numbers)."

Includes bibliographical references.

ISBN 0-8218-2738-3 (alk. paper)

1. Determinantal varieties. 2. Schemes (Algebraic geometry). 3. Liaison theory (Mathematics). I. Kleppe, Jan O. (Jan Oddvar), 1947–. II. Series.

QA3 .A57 no. 732
[QA564]
510 s—dc21
[516.3′5] 2001034317

---

## Memoirs of the American Mathematical Society

This journal is devoted entirely to research in pure and applied mathematics.

**Subscription information.** The 2001 subscription begins with volume 149 and consists of six mailings, each containing one or more numbers. Subscription prices for 2001 are $494 list, $395 institutional member. A late charge of 10% of the subscription price will be imposed on orders received from nonmembers after January 1 of the subscription year. Subscribers outside the United States and India must pay a postage surcharge of $31; subscribers in India must pay a postage surcharge of $43. Expedited delivery to destinations in North America $35; elsewhere $130. Each number may be ordered separately; *please specify number* when ordering an individual number. For prices and titles of recently released numbers, see the New Publications sections of the *Notices of the American Mathematical Society.*

**Back number information.** For back issues see the *AMS Catalog of Publications.*

Subscriptions and orders should be addressed to the American Mathematical Society, P. O. Box 845904, Boston, MA 02284-5904. *All orders must be accompanied by payment.* Other correspondence should be addressed to Box 6248, Providence, RI 02940-6248.



---

*Memoirs of the American Mathematical Society* is published bimonthly (each volume consisting usually of more than one number) by the American Mathematical Society at 201 Charles Street, Providence, RI 02904-2294. Periodicals postage paid at Providence, RI. Postmaster: Send address changes to Memoirs, American Mathematical Society, P. O. Box 6248, Providence, RI 02940-6248.



This publication is indexed in *Science Citation Index®*, *SciSearch®*, *Research Alert®*, *CompuMath Citation Index®*, *Current Contents®/Physical, Chemical & Earth Sciences.*
Printed in the United States of America.

♾ The paper used in this book is acid-free and falls within the guidelines established to ensure permanence and durability.
Visit the AMS home page at URL: `http://www.ams.org/`

10 9 8 7 6 5 4 3 2 1 06 05 04 03 02 01

# Contents

# Abstract

This paper contributes to the liaison and obstruction theory of subschemes in $\mathbb{P}^n$ having codimension at least three.

The first part establishes several basic results on Gorenstein liaison. A classical result of Gaeta on liaison classes of projectively normal curves in $\mathbb{P}^3$ is generalized to the statement that every codimension $c$ "standard determinantal scheme" (i.e. a scheme defined by the maximal minors of a $t \times (t+c-1)$ homogeneous matrix), is in the Gorenstein liaison class of a complete intersection. Then Gorenstein liaison (G-liaison) theory is developed as a theory of generalized divisors on arithmetically Cohen-Macaulay schemes. In particular, a rather general construction of basic double G-linkage is introduced, which preserves the even G-liaison class. This construction extends the notion of basic double linkage, which plays a fundamental role in the codimension two situation.

The second part of the paper studies groups which are invariant under complete intersection linkage, and gives a number of geometric applications of these invariants. Several differences between Gorenstein and complete intersection liaison are highlighted. For example, it turns out that linearly equivalent divisors on a smooth arithmetically Cohen-Macaulay subscheme belong, in general, to different complete intersection liaison classes, but they are always contained in the same even Gorenstein liaison class.

The third part develops the interplay between liaison theory and obstruction theory and includes dimension estimates of various Hilbert schemes. For example, it is shown that most standard determinantal subschemes of codimension 3 are unobstructed, and the dimensions of their components in the corresponding Hilbert schemes are computed.

---

Received by the editor October 9, 1998.

2000 *Mathematics Subject Classification.* Primary 14M12, 14C05, 14H10, 14J15; Secondary 14N05.

*Key words and phrases.* Liaison, linkage, Gorenstein, complete intersection, divisor, canonical module, normal sheaf, liaison invariants, Hilbert scheme, unobstructed schemes, arithmetically Cohen-Macaulay.

The third author was supported in part by DGICYT PB97-0893.

CHAPTER 1

# Introduction

Let $V_1, V_2 \subset \mathbb{P}^n$ be two equidimensional schemes without embedded components. $V_1$ and $V_2$ are said to be directly linked by a complete intersection (directly CI-linked) if there exists a complete intersection scheme $X$ such that $\mathcal{I}_{V_1}/\mathcal{I}_X \cong \mathcal{H}om_{\mathcal{O}_{\mathbb{P}^n}}(\mathcal{O}_{V_2}, \mathcal{O}_X)$ and $\mathcal{I}_{V_2}/\mathcal{I}_X \cong \mathcal{H}om_{\mathcal{O}_{\mathbb{P}^n}}(\mathcal{O}_{V_1}, \mathcal{O}_X)$. If $V_1$ and $V_2$ do not share any common components then this is equivalent to $X = V_1 \cup V_2$. In other words, $V_1$ is directly linked to $V_2$ if their "union" is a complete intersection. The philosophy behind CI-liaison theory is that complete intersections are a well understood class of schemes and that non-complete intersections can be studied by extending them with a complementary scheme to a complete intersection. One studies the complete intersection and the complementary scheme to gain information about the original scheme. A simple example of this is to let $V_1$ be a twisted cubic in $\mathbb{P}^3$ and to let $V_2$ be a secant line to $V_1$. The union of $V_1$ and $V_2$ will be a degree 4 curve which is a complete intersection of two quadric surfaces. By studying the secant line and the complete intersection of the two quadrics one gets information about the twisted cubic. It should be noted that we can reverse this process. Start with a line, $L$, and intersect two general quadric surfaces containing the line to get a degree 4 curve, $C$. The ideal quotient $I_C : I_L$ corresponds to the part of $C$ complementary to the line. This complementary component is a twisted cubic curve. In other words, the complement of a line in a complete intersection of two quadrics containing the line is a twisted cubic curve. A natural way to extend the definition of direct CI-linkage is as follows. Two schemes, $V_1$ and $V_2$, are said to be CI-linked if there exists a sequence of schemes $A_1, \ldots, A_r$ such that $A_i$ is directly CI-linked to $A_{i+1}$ and such that $A_1 = V_1$ and $A_r = V_2$. In this manner, direct CI-linkage generates an equivalence relation called CI-linkage. CI-liaison theory is the study of this equivalence relation and the corresponding equivalence classes.

CI-liaison theory turns out to have a number of rather extraordinary properties. One of the most important of these properties is that the numerical invariants of the schemes $V_1$, $V_2$ and $X$ are all very closely tied together. In fact, one can write a free resolution for any one of the schemes in terms of a free resolution of the other two. In addition, CI-liaison theory is closely intertwined with obstruction theory. We spend a large part of this paper explaining certain CI-liaison invariants and their applications to obstruction theory.

As shown in papers of Schenzel [**66**] and Nagel [**57**], the condition that allows for many of the aforementioned properties to hold is that complete intersections are arithmetically Cohen-Macaulay and have a cyclic dualizing module. Suppose we replace the role of complete intersections in the theory with schemes which have the property of being arithmetically Cohen-Macaulay with cyclic dualizing module. In other words, replace the role played by complete intersections with arithmetically

Gorenstein schemes. Schenzel's and Nagel's work showed that many nice properties are retained. In codimension two, complete intersection schemes and arithmetically Gorenstein schemes coincide but this is definitely not the case in higher codimension. Mimicking the above definitions for CI-liaison theory, we define two schemes, $V_1$ and $V_2$, to be directly linked by an arithmetically Gorenstein scheme (directly G-linked) if there exists an arithmetically Gorenstein scheme $X$ such that $\mathcal{I}_{V_1}/\mathcal{I}_X \cong \mathcal{H}om_{\mathcal{O}_{\mathbb{P}^n}}(\mathcal{O}_{V_2}, \mathcal{O}_X)$ and $\mathcal{I}_{V_2}/\mathcal{I}_X \cong \mathcal{H}om_{\mathcal{O}_{\mathbb{P}^n}}(\mathcal{O}_{V_1}, \mathcal{O}_X)$. If $V_1$ and $V_2$ do not share any common components then this is equivalent to $X = V_1 \cup V_2$. In other words, $V_1$ is directly G-linked to $V_2$ if their "union" is an arithmetically Gorenstein scheme. G-liaison theory is the study of the equivalence relation generated by direct G-linkage and the corresponding equivalence classes.

We note that among subschemes of $\mathbb{P}^n$, this notion cannot be further generalized (in a useful way) by studying Cohen-Macaulay liaison (i.e. allowing $X$ above to be simply arithmetically Cohen-Macaulay). Indeed, Walter [**71**] has shown that *any* unmixed subscheme of $\mathbb{P}^n$ can be linked in this way in a finite number of steps to an arithmetically Cohen-Macaulay scheme; hence there is only one equivalence class in ACM-liaison theory (but see Remark 2.11).

An important question to ask is whether G-liaison theory is a useful equivalence relation to study and can effective computations be made? One of the key points of the present paper is to answer this question with a resounding "yes"! As is to be expected, there are a number of technical obstacles to overcome. Fortunately, the work of Schenzel and Nagel took care of several of these problems but other obstacles remained. These included finding a proper geometric context for the theory, determining methods of completing a given scheme to an arithmetically Gorenstein scheme, characterizing G-liaison classes, etc. We will address these obstacles and provide examples to illustrate various aspects of the theory. See [**52**] for a detailed introduction to liaison theory.

The kernel of the idea of liaison theory is found at least as early as the work of Severi and Macaulay. Max Noether made use of similar ideas in the late nineteenth century. However, the first formal definitions and the first theorem relating specifically to liaison theory appear in the work of Dubreil [**19**] in 1935. In this paper, Dubreil proves that if two curves are directly CI-linked then they have the same projective dimension. A similar statement is definitely not true for surfaces. For instance, the projection of the Veronese surface to $\mathbb{P}^4$ is directly linked by two cubic hypersurfaces to a quintic elliptic scroll. The projection of the Veronese is clearly not projectively normal while the quintic elliptic scroll is projectively normal. In 1945, Apéry [**1**], [**2**] announced that all projectively normal curves in $\mathbb{P}^3$ are in the same liaison class (both CI-liaison class and G-liaison class since the two theories coincide for codimension two). It was not until the paper of Gaeta [**24**] that we find the first formal proof of this fact. When combined with the result of Dubreil, this yields the aesthetic statement that a curve in $\mathbb{P}^3$ is in the linkage class of a complete intersection if and only if the curve is projectively normal. By the Hilbert-Burch Theorem this is equivalent to saying that a curve in $\mathbb{P}^3$ is in the linkage class of a complete intersection if and only if the curve is determinantal (recall that for a curve in $\mathbb{P}^3$ to be determinantal it must be defined by the maximal minors of a $t \times (t+1)$ matrix).

In 1974, Peskine and Szpiro [**62**] published their landmark paper giving a rigorous, modern proof of Gaeta's Theorem and establishing liaison theory as a modern discipline. In 1979, a remarkable paper of Rao [**63**] gave a complete characterization of liaison classes of curves in $\mathbb{P}^3$. He later extended his characterization to the case of locally Cohen-Macaulay codimension two schemes in $\mathbb{P}^n$ [**64**]. A major aspect of these papers was an identification of codimension two liaison classes with stable equivalence classes of vector bundles on $\mathbb{P}^n$ with vanishing first cohomology modules. This allowed one to determine precisely when two schemes fell into the same liaison class (provided they are locally Cohen-Macaulay of codimension two). Shortly after these results were published, Lazarsfeld and Rao [**48**] showed that certain liaison classes of curves in $\mathbb{P}^3$ have a very interesting property. They can be built up entirely from a distinguished element by a process known as basic double linkage and a certain type of cohomologically preserving deformation. Martin-Deschamps and Perrin [**49**] then showed that *every* liaison class of curves in $\mathbb{P}^3$ has this property provided you let the distinguished element be unique up to deformation and they even included an algorithm allowing one to find a distinguished element. Independently, Ballico, Bolondi and Migliore [**4**] showed that the result holds true not just for curves in $\mathbb{P}^3$ but for any codimension two locally Cohen-Macaulay scheme in $\mathbb{P}^n$! The algorithm of Martin-Deschamps and Perrin can be extended to this more general setting. These results and the accompanying algorithm provide a powerful tool for the customized construction of codimension two schemes with prescribed properties and prescribed cohomology. Work of Huneke and Ulrich [**35**], [**36**] placed liaison theory in a wonderfully more general algebraic setting. At the same time, big differences were beginning to show up between CI-liaison theory in codimension two and CI-liaison theory in codimension three. For example, while curves in $\mathbb{P}^4$ have the property that directly linked curves have the same projective dimension, Huneke and Ulrich showed that two projectively normal curves need not be linked! In fact there are an infinite number of projectively normal CI-liaison classes of curves in $\mathbb{P}^4$. One of the simplest examples is that the $2 \times 2$ minors of a $2 \times 4$ matrix of general linear forms in $k[x_0, \ldots, x_4]$ define a curve in $\mathbb{P}^4$ which is projectively normal yet is not in the CI-linkage class of a complete intersection. The thesis of Buchweitz [**13**], later work by Buchweitz and Ulrich [**14**] as well as work by Huneke [**35**] introduced several new CI-liaison invariants and applied them in a number of settings. Their work also established important connections between CI-liaison theory and obstruction theory. Kleppe furthered this connection in a series of papers [**38**]-[**43**]. A large portion of the present paper is devoted to explaining these important invariants, giving applications to obstruction theory and providing sharp formulas for the dimension of certain Hilbert schemes.

An important point of view taken in [**49**] is to view the liaison theory of curves in $\mathbb{P}^3$ as a theory about divisors on surfaces in $\mathbb{P}^3$. For instance, if the two curves, $C_1$ and $C_2$ are linearly equivalent on the surface $F$ in $\mathbb{P}^3$ then they can be shown to be linked in two steps. Furthermore, basic double linkage corresponds naturally to adding hypersurface sections and deformations within a linear series preserve the equivalence class. This point of view can be generalized considerably. Replacing $F$ by any smooth complete intersection scheme, basic double linkage still corresponds to adding hyperplane sections and deformations within linear series still preserve the liaison class. Hartshorne [**30**] showed that, with some care, one can generalize the notion of divisor allowing one to remove the condition of smoothness of $F$. Thus, CI-liaison theory really should be viewed as a theory about divisors on complete intersection schemes! What about G-liaison theory? A first guess might be that G-liaison theory is a theory about divisors on arithmetically Gorenstein schemes but this is a gross underestimate. We will show that G-liaison theory is a theory of divisors on the very large class of arithmetically Cohen-Macaulay schemes. This places a formerly very algebraic theory into a very geometric context. We spend some time applying this point of view to a number of problems then collect some of the remaining open questions in the final chapter.

Chapter 1 and Chapter 2 of this paper provide (respectively) background and preliminary results. As mentioned earlier, Gaeta proved that every projectively normal curve (or equivalently every determinantal curve) in $\mathbb{P}^3$ is linked to a complete intersection. In the context of CI-liaison, Gaeta's Theorem does not generalize well to higher codimension. As mentioned earlier, the determinantal curve defined by the maximal minors of a $2 \times 4$ matrix of general linear forms in $k[x_0, \dots x_4]$ is not CI-linked to a complete intersection. In the context of G-liaison, a generalization of Gaeta's Theorem does hold. A scheme, $X$, of codimension $c+1$ is said to be *standard determinantal* if its homogeneous ideal can be generated by the maximal minors of a homogeneous $t \times (t+c)$ matrix. Most of Chapter 3 is devoted to proving the following result.

THEOREM 1.1. *Every standard determinantal scheme in $\mathbb{P}^n$ is Gorenstein linked to a complete intersection.*

We were pleasantly surprised that the result generalized so cleanly to arbitrary codimension.

The first half of Chapter 4 begins the development of G-liaison theory as a theory of divisors on arithmetically Cohen-Macaulay schemes. To put the theory into as broad a context as possible, we utilize the ideas of Hartshorne on generalized divisors [**30**]. The second half of Chapter 4 is concerned with the equivalence of algebraic CI-linkage and geometric CI-linkage, extending work of Rao [**63**] for codimension two. Letting $H(c,n)$ denote the set of all subschemes of $\mathbb{P}^n$ that are equidimensional of codimension $c$ and generic complete intersections, we show

THEOREM 1.2. *Algebraic and geometric CI-linkage generate the same equivalence relation on $H(c,n)$. That is, if $C, C' \in H(c,n)$ and there is a sequence of algebraic CI-links $C \sim C_1 \sim \cdots \sim C_t \sim C'$ with each $C_i \in H(c,n)$ then there is a sequence of geometric CI-links from $C$ to $C'$.*

In the special case of codimension two, we improve this theorem, removing the condition that each $C_i \in H(c,n)$, extending work of Schwartau for curves in $\mathbb{P}^3$ [**67**].

Chapter 5 continues the development of G-liaison theory as a theory of divisors on arithmetically Cohen-Macaulay schemes. When the arithmetically Cohen-Macaulay scheme happens to be a complete intersection, the theory collapses to the setting of CI-liaison theory as a theory of divisors on a complete intersection. First, it is shown that twisted anticanonical divisors on arithmetically Cohen-Macaulay schemes which satisfy property $G_1$ (see Definition 2.1) are arithmetically Gorenstein. While this is not so difficult to prove, it is quite central to much of the development of the theory. It provides an effective method of extending a given scheme to an arithmetically Gorenstein scheme. In fact, it allows us to prove the following very useful result:

PROPOSITION 1.3. *Let $S$ be an arithmetically Cohen-Macaulay subscheme of $\mathbb{P}^n$. Let $C$ be a closed subscheme of $S$ of codimension one in $S$. Let $F$ be a* *homogeneous polynomial of degree $t$ not vanishing on any component of $S$. Then we have:*

(i) *The ideal $I_S + F \cdot I_C$ is the saturated ideal of an arithmetically Cohen-Macaulay subscheme $Y$ of $S$. If $F$ is sufficiently general then $Y$ is the union of $C$ with the hypersurface section $H_F$ of $S$ by $F$.*

(ii) *If $S$ has property $G_0$ (generically Gorenstein) then $Y$ is G-linked to $C$ in two steps; hence we call $Y$ a* Basic Double G-link *of $C$.*

(iii) *If $S$ has property $G_1$ then the linear equivalence class of $Y$ makes sense, and we have that any divisor in $|C + tH|$ is G-linked to $Y$ in two steps.*

For CI-liaison, the above fact is true if $S$ is a complete intersection. It should be noted that linearly equivalent divisors on non complete intersections are not, in general, CI-linked. Compare this, for instance, with Proposition 1.5 below.

In Chapter 6, we consider groups which are invariant under (algebraic) CI-linkage. It is well known that for a (locally) Cohen-Macaulay scheme $X$ of dimension $d$ in $\mathbb{P}^n$, the cohomology groups $H^i_*(\mathbb{P}^n, \mathcal{I}_X)$ (the Rao invariants) are invariant up to shifts and duals if $1 \le i \le d$ (even for G-liaison [**66**]). $X$ is arithmetically Cohen-Macaulay if and only if all of these groups vanish. Buchweitz and Ulrich proved the CI-liaison invariance of $H^i_{\mathfrak{m}}(K_{R/I_X} \otimes_R I_X)$ and of most $H^i_*(\mathcal{N}_X)$ in an unpublished manuscript [**14**]. These invariants are important in that they allow one to distinguish between many CI-liaison classes which cannot be distinguished by the Rao invariants alone. We give a geometric approach to the proof of this fact (extending the approach originating in [**37**], [**38**]). We prove

THEOREM 1.4. *Let $X \subset \mathbb{P}^N$ and $X' \subset \mathbb{P}^N$ (with saturated ideals $I_X$ and $I_{X'}$ of $R$ resp.) be ACM schemes, algebraically linked by some complete intersection $Y \subset \mathbb{P}^N$ of dimension $n$. Let $\mathcal{N}_X$ be the normal sheaf of $X$ and consider the natural map*

$$\delta_X : H^n_*(\mathcal{N}_X) \to \operatorname{Hom}_R(I_X, H^{n+1}_{\mathfrak{m}}(R/I_X)).$$

*If $n > 0$, then, as graded $R$-modules we have*

$$H^i_*(\mathcal{N}_X) \cong H^i_*(\mathcal{N}_{X'}) \quad \text{for } 1 \le i \le n-1, \qquad \text{and}$$

$$\ker \delta_X \cong \ker \delta_{X'}.$$

*Moreover, at least as graded $k$-modules, we have for any $n \ge 0$*

$$\operatorname{coker} \delta_X \cong \operatorname{coker} \delta_{X'}.$$

We give a number of applications of this theorem throughout Chapter 6 and the rest of the paper.

Chapter 7 is concerned with geometric applications of the ideas in Chapter 6. We study the behavior of these CI-liaison invariants under the operation of basic double linkage on an arithmetically Cohen-Macaulay scheme. In other words, we compare the invariants for a divisor $C \subset S$ to the invariants for the divisor $C + tH$ when $S$ is an arithmetically Cohen-Macaulay scheme. This leads us to the following result (among others).

PROPOSITION 1.5. *Let $C \subset S$ be two ACM schemes, $n = \dim C \ge 0$, where $S$ is generically, but not globally, a complete intersection in $\mathbb{P}^{n+3}$. Moreover, suppose $C$ is a Cartier divisor on $S$ satisfying $\mathcal{O}_S(C) \cong \omega_S(h)$ for some $h$, and let $C_v \in$*

*$|C + vH|$ be effective Cartier divisors. Then $C_v$ and $C_{v'}$ are not licci and they belong to different CI-liaison classes for any $v > v' \gg 0$.*

In Chapter 7 we moreover introduce a graded module, $L^0(C)$, and we show that the conclusion of Proposition 1.5 holds provided $L^0(C)_v \neq 0$ for some integer $v$, making the proposition above a very special case. In this manner we can construct infinitely many CI-liaison classes of arithmetically Cohen-Macaulay schemes lying on a given arithmetically Cohen-Macaulay scheme. CI-liaison theory does not behave well on such a scheme.

Chapter 8 demonstrates how G-liaison improves the situation considerably. We had hoped to resolve the following question.

QUESTION 1.6. *In any projective space, is there only one Gorenstein liaison class containing ACM schemes of fixed codimension?*

This should be viewed as a further generalization of Gaeta's theorem, beyond our result on standard determinantal schemes which we prove in Chapter 3. While this may be a difficult question, we did make substantial progress toward an answer. For instance, we established the following rather general result.

PROPOSITION 1.7. *Let $X$ be a smooth, ACM (hence integral) surface in $\mathbb{P}^n$ and let $M$ be a graded $k[X_0, ..., X_n]$-module of finite length. Then there exist only finitely many G-liaison classes of curves $C \subset X$ with Rao module $M(C) \cong M$ (up to twist).*

Setting $M = 0$ shows that there are only finitely many liaison classes of projectively normal curves on a given smooth arithmetically Cohen-Macaulay surface. Furthermore, in a case by case analysis we show that

PROPOSITION 1.8. *All ACM curves $C \subset \mathbb{P}^4$ lying on a general smooth, rational ACM surface $X \subset \mathbb{P}^4$ are glicci.*

Further progress on Question 1.6 has been made recently by the third author with M. Casanellas [**15**], [**16**], in which they extend the above proposition to a much larger class of arithmetically Cohen-Macaulay surfaces in $\mathbb{P}^4$, and to a much more general setting of divisors in higher codimension. In a different direction, the second and fourth authors have recently shown [**53**] that any arithmetically Cohen-Macaulay subscheme defined by a Borel-fixed monomial ideal is glicci, and also that for any Hilbert function which exists for arithmetically Cohen-Macaulay subschemes, a reduced scheme with that Hilbert function can be found which is glicci.

Chapter 9 is concerned with the unobstructedness and the dimension of certain families of subschemes of $\mathbb{P}^n$ obtained by adding hypersurface sections and with the unobstructedness of most determinantal schemes. In earlier sections it was shown that adding hypersurface sections of high enough degree results in schemes lying in different CI-liaison classes but the same G-liaison class. It was also shown that determinantal schemes are all G-linked to a complete intersection but are rarely CI-linked to a complete intersection. As seen in [**38**] unobstructedness is preserved under CI-linkage. Since all determinantal schemes are G-linked to a complete intersection and since there exist obstructed determinantal schemes one wonders if G-liaison ever preserves unobstructedness. We show that the answer is "yes" under certain conditions. The first theorem of Chapter 9 is

THEOREM 1.9. *Let $C \subset S$ be equidimensional, locally Cohen-Macaulay closed subschemes of $\mathbb{P}^n$ such that $C$ is a Cartier divisor on $S$. Suppose that $S$ is unobstructed and satisfies $H^2(R, R/I_S, R/I_S)^\sim = 0$. Let $t$ be an integer such that $H^1(\mathcal{O}_S(C)(t)) = 0$, and suppose the relative Picard scheme Pic is formally smooth (see Remark 9.1) at $(\mathcal{O}_S(C), S)$. If $C_t \in |C + tH|$ is an arbitrary effective Cartier divisor inducing an injective map $H^1(\mathcal{N}_S \otimes \mathcal{I}_{C|S}(-t)) \to H^1(\mathcal{N}_S)$, then $C_t$ is unobstructed, and we have*

$$\begin{aligned}\dim_{C_t} \mathrm{Hilb}^p(\mathbb{P}^N) &= h^0(\mathcal{N}_S) + h^1(\mathcal{O}_S) - \dim Im\, \alpha_L - h^0(\mathcal{N}_S \otimes \mathcal{I}_{C|S}(-t)) \\ &\quad + h^0(\mathcal{O}_S(C)(t)) - 1,\end{aligned}$$

*where $\alpha_L$ is a certain map defined on page 81.*

The proof of this theorem uses the theory of Hilbert flag schemes. We use this theorem to derive a number of results on obstructedness as well as to make explicit formulas for the dimension of Hilbert schemes of several large classes of schemes. For example we show

PROPOSITION 1.10. *Let $C \subset \mathbb{P}^4$ be a smooth, ACM curve lying on a smooth, rational ACM surface $S \subset \mathbb{P}^4$. If $\deg(C) \leq 8$ or $\deg(C) \geq 24$ then $C$ is unobstructed.*

The latter part of Chapter 9 is concerned with dimension formulas and unobstructedness of schemes in the strata $\mathrm{GradAlg}^H$ of $\mathrm{Hilb}^p$, where we deform $C = \mathrm{Proj}(A) \subset \mathbb{P}^N$ with constant Hilbert function $H = H_A = H_C$. $\mathrm{GradAlg}^H$ allows a natural scheme structure [**41**] whose tangent space is ${}_0H^1(R, A, A)$ (resp. ${}_0H^2(R, A, A)$). $C$ will be called *unobstructed in* $\mathrm{GradAlg}^H$ (or *$H$-unobstructed*) if $\mathrm{GradAlg}^H$ is smooth at $(C \subset \mathbb{P}^N)$. We show

THEOREM 1.11. *Let $C$ be a Cartier divisor on $S \subset \mathbb{P}^n$, where $S$ is an $H_S$-unobstructed curve (so $C$ is a zeroscheme). Let $C_t \in |C + tH|$ be an arbitrary effective Cartier divisor such that ${}_0\mathrm{Hom}_R(I_{C_t|S}, H^2_{\mathfrak{m}}(I_{C_t})) = 0$. If*

$${}_0\mathrm{Ext}^1_R(I_S, I_{C_t|S}) \to {}_0\mathrm{Ext}^1_R(I_S, R/I_S)$$

*is injective, then $C_t$ is $H_{C_t}$-unobstructed. If moreover $S$ is an ACM curve and a local complete intersection along $C_t$, then $C_t$ is unobstructed and*

$$\begin{aligned}&\dim_{C_t} \mathrm{GradAlg}^{H_{C_t}} = h^0(\mathcal{N}_{C_t|S}) + h^0(\mathcal{N}_S) - h^0(\mathcal{N}_S \otimes \mathcal{I}_{C|S}(-t)),\\ &\dim_{C_t} \mathrm{Hilb}^{p_t} - \dim_{C_t} \mathrm{GradAlg}^{H_{C_t}} = h^1(\mathcal{N}_S \otimes \mathcal{I}_{C|S}(-t)) - h^1(\mathcal{N}_S)\end{aligned}$$

*where $h^0(\mathcal{N}_{C_t|S}) = \chi(\mathcal{O}_S(C)(t)) - \chi(\mathcal{O}_S)$.*

In Chapter 10 we give an upper bound for the dimension of the locus $W(\underline{b}, \underline{a})$ of good determinantal schemes $C \subset \mathbb{P}^{n+c}$ (cf. Definition 3.1). The bound is sharp in a number of instances. The main result of Chapter 10 (Theorem 10.13) is somewhat technical, although the conditions of the theorem can be shown to be satisfied in a wide number of cases which we make explicit in Chapter 10. For instance, we have the following result.

PROPOSITION 1.12. *Let $C \subset S$ be good determinantal schemes in $\mathbb{P}^{n+3}$ where $C$ is given by the maximal minors of a $t \times (t + 2)$ matrix whose $ij$-th entry is homogeneous of degree $a_j - b_i$, $0 \leq j \leq t + 1$, $1 \leq i \leq t$, and $S$ is correspondingly* *given by deleting some column. Let*

$$l = \sum_{j=0}^{t+1} a_j - \sum_{i=1}^{t} b_i, \quad a_r = \max\{a_j\}$$

*and suppose* $C \subset S$ *is Cartier in codimension* $\leq 3$ *in* $S$, $S$ *is* $G_1$, *and* $n = \dim C \geq 1$. *Then*

(i) $\dim W(\underline{b};\underline{a}) = \lambda + \kappa$ *where*

$$\begin{aligned}
\lambda &= \sum_{i,j} \binom{a_j - b_i + n + 3}{n+3} + \sum_{i,j} \binom{b_i - a_j + n + 3}{n+3} - \sum_{i,j} \binom{b_i - b_j + n + 3}{n+3} \\
&\quad - \sum_{i,j} \binom{a_j - a_i + n + 3}{n+3} + 1 \\
\kappa &= \binom{-l + 2a_r + n + 3}{n+3}.
\end{aligned}$$

(ii) *Suppose* $\dim C > 1$ *or if* $\dim C = 1$, *we suppose* $H^1(\mathcal{I}_S/\mathcal{I}_S^2(l - 2a_k)) = 0$ *provided* $S$ *is obtained deleting the* $(k+1)$*-th column. Then* $C$ *is unobstructed, and*

$$\dim_C \operatorname{Hilb}^p(\mathbb{P}^{n+3}) = \dim W(\underline{b};\underline{a}) = \lambda + \kappa.$$

We would like to thank the referee for his or her useful suggestions and comments, and especially for the ideas to simplify the first part of the proof of Proposition 6.8. We would also like to thank Robin Hartshorne for his careful checking of Chapter 8, which led to Remark 8.9 and a revision of the proof of Corollary 8.10.

CHAPTER 2

# Preliminaries

Throughout this paper, $R$ will denote the polynomial ring $R = k[x_0, x_1, \dots, x_n]$ where $k$ is a field of characteristic zero, and $\mathbb{P}^n = \mathrm{Proj}(R)$. Starting in Chapter 6, we will assume that $k$ is algebraically closed. We will denote by $\mathfrak{m}$ the maximal ideal in $R$. The sheafification of a module $M$ will be denoted by $M^{\sim}$. $\mathcal{I}_X$ (resp. $\mathcal{N}_X$) is the ideal sheaf (resp. normal sheaf) of a closed subscheme $X$ of $\mathbb{P}^n$, and $\mathcal{I}_{X|Y}$ is the ideal sheaf of $X$ in $Y$. We will use $I_X$ to denote the saturated ideal $H^0_*(\mathbb{P}^n, \mathcal{I}_X)$ and $A$ will be used to denote $R/I_X$. When there is no danger of ambiguity we sometimes simply write $I$ for the homogeneous ideal of $X$. For a homogeneous polynomial $F \in R$, the principal ideal $F \cdot R$ we will denote by $(F)$. We use $M^\vee$ to denote the dual of $M$ as a $k$-vector space (resp. the graded $k$-dual of $M$ if $M$ is graded) and $\mathrm{rank}_k M$ to denote the rank of $M$ as a $k$-vector space. The ring $B = R/I_Y$ is usually a complete intersection in $R$ and $(-)^*$ denotes its $B$-dual (or its $\mathcal{O}_Y$-dual where $Y = \mathrm{Proj}(B)$). If $X = \mathrm{Proj}(R/I) \subset \mathbb{P}^n = \mathrm{Proj}(R)$ is a closed subscheme of codimension $c$, and $A = R/I$, we denote by $K_A$ (or sometimes by $K_X$) its canonical module $\mathrm{Ext}^c_R(A, R)(-n-1)$ and by $\omega_X$ its canonical sheaf $\mathcal{E}xt^c_{\mathcal{O}_{\mathbb{P}^n}}(\mathcal{O}_X, \mathcal{O}_{\mathbb{P}^n})(-n-1)$.

DEFINITION 2.1. A noetherian ring $A$ (resp. a noetherian scheme $X$) satisfies the condition $G_r$, "Gorenstein in codimension $\leq r$" if every localization $A_{\mathfrak{p}}$ (resp. every local ring $\mathcal{O}_x$) of dimension $\leq r$ is a Gorenstein local ring.

REMARK 2.2. Definition 2.1 means that the non locally Gorenstein locus has codimension greater than $r$. In particular, $G_0$ is "generically Gorenstein." See **[30]** for more details on schemes satisfying the condition $G_r$.

DEFINITION 2.3. Let $V_1$ and $V_2$ be two non-empty equidimensional schemes without embedded components. $V_1$ and $V_2$ are said to be *directly CI-linked* (resp. *directly G-linked*) if there exists a complete intersection scheme (resp. arithmetically Gorenstein scheme) $X$ such that

$$\mathcal{I}_{V_1}/\mathcal{I}_X \cong \mathcal{H}om_{\mathcal{O}_{\mathbb{P}^n}}(\mathcal{O}_{V_2}, \mathcal{O}_X) \quad \text{and} \quad \mathcal{I}_{V_2}/\mathcal{I}_X \cong \mathcal{H}om_{\mathcal{O}_{\mathbb{P}^n}}(\mathcal{O}_{V_1}, \mathcal{O}_X).$$

DEFINITION 2.4. Two schemes, $V_1, V_2 \subset \mathbb{P}^n$, are said to be *CI-linked* (resp. *G-linked*) if there exists a sequence of schemes $A_1, \dots, A_r$ such that $A_i$ is directly CI-linked (resp. directly G-linked) to $A_{i+1}$ and such that $A_1 = V_1$ and $A_r = V_2$. We say that $V_1$ and $V_2$ are *CI-bilinked* (resp. *G-bilinked*) if $r = 3$. In other words, if $V_1$ is linked to $V_2$ in 2 steps.

Complete intersection liaison (or CI-liaison) is the study of the equivalence relation determined by CI-linkage. Gorenstein liaison (or G-liaison) is the study of the equivalence relation determined by G-linkage. One of the purposes of this paper is to show that Gorenstein liaison is a very natural geometric theory of divisors on an arithmetically Cohen-Macaulay scheme. We begin with the observation that the standard formula relating the degrees and arithmetic genera of *curves* which are directly linked by a complete intersection (cf. [**62**] Proposition 3.1, [**30**] Remark 4.7.1 and [**52**]) is in fact a special case of a very simple, more general formula for G-linked curves. The following proposition is a direct consequence of [**57**], Corollary 3.6, which in fact computes the behavior of the Hilbert function under G-liaison (as is done also in [**30**] for complete intersection liaison). See also [**56**] pp. 550–551. However, this formulation seems to be new, and we outline a simple proof. See [**52**] for more details.

PROPOSITION 2.5. *Let $X$ be an arithmetically Gorenstein curve in $\mathbb{P}^n$ with minimal free resolution*

$$0 \to R(-a) \to \cdots \to I_X \to 0.$$

*Then*

(i) $p_a(X) = \frac{1}{2}(a-n-1)(deg\ X) + 1$;

(ii) *Suppose that $C_1$ and $C_2$ are curves which are directly G-linked by $X$. Then*

$$p_a(C_1) - p_a(C_2) = \frac{1}{2}(a-n-1)(deg\ C_1 - deg\ C_2).$$

PROOF. Because $X$ is arithmetically Gorenstein, we have $\omega_X \cong \mathcal{O}_X(a-n-1)$. Using this fact and standard Riemann-Roch techniques one can show that if $K$ is a canonical divisor on $X$ then $\deg K = 2p_a(X) - 2$. That is,

$$2p_a(X) - 2 = \deg K = (a-n-1)(\deg X).$$

This gives part (i).

For (ii), we have the exact sequence

$$0 \to I_X \to I_{C_1} \to H^0_*(C_2, \omega_{C_2})(n+1-a) \to 0.$$

Using this sequence, a Hilbert polynomial calculation then gives the formula

$$p_a(X) - 1 - (\deg C_2)(a-n-1) = p_a(C_1) - p_a(C_2).$$

Using (i) and the fact that $\deg X = \deg C_1 + \deg C_2$ gives (ii). □

REMARK 2.6. In Proposition 2.5, if $G$ is the complete intersection of hypersurfaces of degrees $d_1, \dots, d_{n-1}$ then $a = \sum d_i$ and $\deg G = \prod d_i$, and so (i) and (ii) are generalizations of standard formulas for complete intersections. Notice, for example, that it continues to hold, as is well-known in the complete intersection case, that if $C_1$ and $C_2$ have the same degree then they necessarily have the same arithmetic genus.

DEFINITION 2.7. Given a scheme $X \subset \mathbb{P}^n$ with ideal sheaf $\mathcal{I}_X$ the $i^{th}$ *Rao module* (or $i^{th}$ *deficiency module*) of $X$ is the module $M_i(X) = H^i_*(\mathbb{P}^n, \mathcal{I}_X)$. If $X$ is locally Cohen-Macaulay and equidimensional of codimension $c$ then $M_i(X)$ is a finite length module for $1 \leq i \leq n-c$.

DEFINITION 2.8. Assume the scheme $X \subset \mathbb{P}^n$ is locally Cohen-Macaulay and equidimensional of codimension $c$. The scheme $X$ is said to be *arithmetically Cohen-Macaulay* or ACM if $M_i(X) = 0$ for $1 \leq i \leq n-c$.

REMARK 2.9. An equidimensional curve, $C$, is arithmetically Cohen-Macaulay if and only if it is projectively normal.

DEFINITION 2.10. Following [**36**], we say that a scheme is *licci* if it is in the CI-liaison class of a complete intersection. We say that a scheme is *glicci* if it is in the G-liaison class of a complete intersection.

REMARK 2.11. While it is the case that schemes which are licci or glicci are necessarily arithmetically Cohen-Macaulay, it is not true that arithmetically Cohen-Macaulay schemes are necessarily licci. On the other hand, it is an open problem whether schemes which are arithmetically Cohen-Macaulay are necessarily glicci, and we discuss this question elsewhere in this paper.

It is interesting to note that many authors do not explicitly require in the definition of linkage (Definitions 2.3 and 2.4) that the schemes in question are all non-empty. Allowing the empty set (i.e. the scheme defined by the saturated ideal consisting of the whole ring) forces us to adopt the unpleasant convention that the empty set has all dimensions, since it is directly "linked" to a complete intersection of any dimension.

One would like to avoid this. For the study of licci schemes, it is a classical result (cf. for instance [**67**]) that any two complete intersections of the same dimension are linked, without recourse to the empty set. On the same note, it was pointed out to the second author by H. Martin that in [**71**], C. Walter showed only that an arbitrary equidimensional subscheme of $\mathbb{P}^n$ is Cohen-Macaulay linked (CM-linked) to an arithmetically Cohen-Macaulay subscheme of $\mathbb{P}^n$, leaving open the question of whether one can pass from an arbitrary Cohen-Macaulay scheme $V$ to another one $V'$ of the same dimension, without recourse to the empty set.

We would like to point out that the answer to this question is "yes." Indeed, one can simply choose a complete intersection $S$ containing $V$ and of dimension one greater, and perform a general basic double link (cf. [**48**], [**25**] or [**7**]). Then the resulting scheme is

- the union of $V$ and a hypersurface section of $S$ (which is a complete intersection), and
- CI-linked to $V$ in two steps, hence arithmetically Cohen-Macaulay.

That is, $V$ is CM-linked to a complete intersection, and similarly we can CM-link $V'$ to another complete intersection. By the above we can find a series of CI-links between these two complete intersections, and all of this avoids the empty set.

There is one interesting question remaining for G-linkage. Is it true that any arithmetically Gorenstein subscheme of $\mathbb{P}^n$ is G-linked to any other arithmetically Gorenstein subscheme of the same dimension (without recourse to the empty set)? This is known to be true in codimension three by a result of Watanabe [**73**] (and in fact a codimension three arithmetically Gorenstein scheme is even licci).

EXAMPLE 2.12. Although our goal in this paper is to show the merits of studying Gorenstein liaison, it is worth mentioning one disadvantage. It is well-known (see [**52**] Proposition 5.2.25) that if $V \subset \mathbb{P}^n$ is any ACM subscheme and $Z \subset \mathbb{P}^{n-1}$ is a general hyperplane section, and if $X$ is a complete intersection in $\mathbb{P}^{n-1}$ CI-linking $Z$ to a residual subscheme $Z'$, then $X$ lifts to a complete intersection in $\mathbb{P}^n$ which CI-links $V$ to a residual subscheme $V'$, and $Z'$ is the hyperplane section of $V'$.

On the other hand, it was shown in [**52**] Example 5.2.26 that if $V$ is a rational normal curve in $\mathbb{P}^4$ and $Z \subset \mathbb{P}^3$ is a general hyperplane section then there is an arithmetically Gorenstein zeroscheme $X$ in $\mathbb{P}^3$ G-linking $Z$ to a single point, such that $X$ does not lift to an arithmetically Gorenstein curve in $\mathbb{P}^4$ containing $V$.

CHAPTER 3

# Gaeta's Theorem

In this chapter we do not assume characteristic zero. It is a classical result, originally proved by Gaeta and re-proved in modern language by Peskine and Szpiro [**62**], that every ACM, codimension two subscheme of $\mathbb{P}^n$ can be linked in a finite number of steps to a complete intersection. (Recall that in codimension two the arithmetically Gorenstein subschemes and the complete intersections coincide, so there is no ambiguity about this statement.) Recall also that the homogeneous ideal of a codimension two ACM subscheme of $\mathbb{P}^n$ is given by the maximal minors of a $t \times (t+1)$ homogeneous matrix, the Hilbert-Burch matrix. That is, such a scheme is *standard determinantal* (see below). The purpose of this chapter is to extend Gaeta's Theorem, viewed as a statement on standard determinantal schemes of codimension two, to arbitrary codimension.

DEFINITION 3.1. If $A$ is a homogeneous matrix, we denote by $I(A)$ the ideal of maximal minors of $A$. If $\varphi : F \to G$ is a homomorphism of free graded $R$-modules then we define $I(\varphi) = I(A)$ for any homogeneous matrix $A$ representing $\varphi$ after a choice of basis for $F$ and $G$. A codimension $c+1$ scheme, $X$, in $\mathbb{P}^n$ will be called a *standard determinantal scheme* if $I_X = I(A)$ for some homogeneous $t \times (t+c)$ matrix, $A$. $X$ will be called a *good determinantal scheme* if additionally, $A$ contains a $(t-1) \times (t+c)$ submatrix (allowing a change of basis if necessary) whose ideal of maximal minors defines a scheme of codimension $c+2$. In a similar way we define standard and good determinantal ideals.

We need some preparations. Extensive information on determinantal ideals can be found in [**11**].

PROPOSITION 3.2. *Let $X \subset \mathbb{P}^n$ be a standard determinantal scheme such that $I_X = I(A)$ for some homogeneous $t \times (t+c)$ matrix $A$. Suppose that the ground field $k$ is infinite. Then the following conditions are equivalent:*

(i) *$X$ is good determinantal.*
(ii) *$X$ is standard determinantal and a generic complete intersection.*
(iii) $\operatorname{codim} I_{t-1}(A) \geq c+2$.

PROOF. The equivalence of (i) and (ii) follows by [**45**], Theorem 3.4 (cf. also Remark 3.5).

Furthermore (i) implies clearly (iii). Thus it suffices to show that (ii) is a consequence of (iii). Let $\varphi : F \to G$ be the homomorphism of free modules defined by $A$. Put $M_\varphi = \operatorname{coker} \varphi$. Let $\mathfrak{p}$ be a minimal prime ideal of $I_X$. Then $\mathfrak{p}$ has codimension $c$ and thus does not contain $I_{t-1}(\varphi)$. Therefore it follows by [**11**], Proposition 16.3 that $(M_\varphi)_\mathfrak{p}$ has at most one minimal generator as $R_\mathfrak{p}$-module. But $(I_X)_\mathfrak{p}$ is generated by the maximal minors of $A_\mathfrak{p}$. Hence we can conclude that $(M_\varphi)_\mathfrak{p}$ has precisely one minimal generator. This means that $\varphi_\mathfrak{p}$ is the sum of an isomorphism of free modules of rank $t-1$ and a map into $R_{\mathfrak{p}}$. Thus $(I_X)_{\mathfrak{p}}$ is a complete intersection as claimed in (ii). □

Now we want to discuss some complexes related to standard determinantal schemes. For more details we refer to [**11**], [**21**] and [**54**]. Let $\varphi : F \to G$ be a homomorphism of free graded $R$-modules of rank $t+c-1$ and $t$, respectively. Then there are (generalized) Koszul complexes $\mathcal{C}_i(\varphi)$:

$$0 \to \wedge^i F \otimes S_0(G) \to \wedge^{i-1} F \otimes S_1(G) \to \ldots \to \wedge^0 F \otimes S_i(G) \to 0.$$

Let $\mathcal{C}_i(\varphi)^*$ be the $R$-dual of $\mathcal{C}_i(\varphi)$. The dual map $\varphi^*$ induces graded homomorphisms

$$\nu_i : \wedge^{t+i} F \otimes \wedge^t G^* \to \wedge^i F.$$

They can be used in order to splice the complexes $\mathcal{C}_{c-1-i}(\varphi)^* \otimes \wedge^{t+c-1} F \otimes \wedge^t G^*$ and $\mathcal{C}_i(\varphi)$ to a complex $\mathcal{D}_i(\varphi)$:

$$\begin{aligned} 0 \to \wedge^{t+c-1} F \otimes S_{c-1-i}(G)^* \otimes \wedge^t G^* &\to \wedge^{t+c-2} F \otimes S_{c-2-i}(G)^* \otimes \wedge^t G^* \to \ldots \\ \to \wedge^{t+i} F \otimes S_0(G)^* \otimes \wedge^t G^* &\xrightarrow{\nu_i} \wedge^i F \otimes S_0(G) \to \wedge^{i-1} F \otimes S_1(G) \to \ldots \\ &\to \wedge^0 F \otimes S_i(G) \to 0. \end{aligned}$$

The complex $\mathcal{D}_0(\varphi)$ is called the Eagon-Northcott complex and $\mathcal{D}_1(\varphi)$ is called the Buchsbaum-Rim complex.

Let us rename the Koszul complex $\mathcal{C}_c(\varphi)$ as $\mathcal{D}_c(\varphi)$. Then we have the following resolutions (cf. [**21**], Theorem A2.10).

PROPOSITION 3.3. *Suppose $I(\varphi)$ defines a standard determinantal scheme $S \subset \mathbb{P}^n$ of codimension $c$. Then it holds:*

(i) *$\mathcal{D}_i(\varphi)$ is acyclic for $0 \le i \le c$.*
(ii) *If $\varphi$ is a minimal homomorphism, i.e. $\operatorname{im} \varphi \subset \mathfrak{m} \cdot G$, then $\mathcal{D}_0(\varphi)$ is a minimal free graded resolution of $R/I_S$ and $\mathcal{D}_i(\varphi)$ is a minimal free graded resolution of $S_i(\operatorname{coker} \varphi)$, $1 \le i \le c$.*

If we have even stronger genericity conditions on $\varphi$ we can identify the symmetric powers as ideals of the homogeneous coordinate ring of $S$.

PROPOSITION 3.4. *Let $\varphi : F \to G$ be a homomorphism between free modules of rank $t+c-1$ and $t$ respectively defining a standard determinantal scheme $S$. Fix bases for $F$ and $G$ such that $\varphi$ is given by a $t \times (t+c-1)$ matrix $A$. Let $A_1$ be the submatrix of $A$ consisting of its first $t-1$ columns and put $J = I(A_1)$. Assume that $I_S + J$ has codimension $\ge c+1$. Then it holds:*

(i) *For $i = 1, \ldots, c$ we have for the $i$-th symmetric power of the cokernel $M_\varphi = \operatorname{coker} \varphi$ that $S_i(M_\varphi) \cong [(I_S + J^i)/I_S](ia)$ where $a$ is the integer defined by $\wedge^{t-1} F'' \otimes R(a) \cong \wedge^t G$ and $F''$ is the direct summand of $F$ such that $A_1$ defines the homomorphism $F'' \to G$.*
(ii) *The isomorphism $M_\varphi \cong [(I_S+J)/I_S](a)$ in* (i) *is given by the homomorphism which maps $y \in G$ (considered as a column) to the residue class in $R/I_S$ of the determinant of the concatenation of $A_1$ and $y$.*
(iii) *If $S$ is good determinantal then we may assume (possibly after a change of bases) that $A_1$ satisfies the above assumption.*

PROOF. Claim (i) is Theorem A2.14 in [**21**]. Its proof shows also the second assertion.

If $S$ is good determinantal then it holds in particular $\operatorname{codim} I_{t-1}(\varphi) \ge c+1$. Thus there is a $(t-1)$-minor $g$ of $A$ such that $I_S : g = I_S$. We may assume that the first $(t-1)$ columns of $A$ are involved in order to get $g$. It follows $I_S + gR \subset I_S + J$. Therefore we see that $\operatorname{codim}(I_S + J) > \operatorname{codim} S$ proving the third claim. □

COROLLARY 3.5. *With the assumptions of Proposition 3.4 it holds:*

(i) *For $i = 1, \ldots, c$ we have that the ideals $I_S + J^i$ are perfect and have codimension $c + 1$.*

(ii) *$I_S + J^{c-1}$ is a Gorenstein ideal.*

PROOF. According to Proposition 3.4 there are exact sequences

$$0 \to S_i(M_\varphi)(-ia) \to R/I_S \to R/(I_S + J^i) \to 0, \quad 1 \leq i \leq c.$$

The module $S_i(M_\varphi)(-ia)$ is a Cohen-Macaulay module of dimension $n + 1 - c$ due to [**21**], Corollary A2.13. The same is true for $R/I_S$ by assumption. It follows that $R/(I_S + J^i)$ is Cohen-Macaulay of dimension $n - c$ proving the first claim.

Now we consider the exact sequence above with $i = c - 1$. A free resolution of $S_{c-1}(M_\varphi)$ is given by the complex $\mathcal{D}_{c-1}(\varphi)$. It has length $c$ and ends with a free module of rank one. Therefore the mapping cone procedure shows that $I_S + J^{c-1}$ has Cohen-Macaulay type one, i.e., it is a Gorenstein ideal. □

Since $S_{c-1}(M_\varphi)$ is up to a degree shift the canonical module of $S$ (cf., for example, [**11**], Theorem 2.20) part (ii) of the preceding statement can be viewed as an initial case of the more general Lemma 5.4 which we will prove later on.

Now we are ready for the main result of this chapter.

THEOREM 3.6. *Every standard determinantal scheme is glicci.*

PROOF. Let $V \subset \mathbb{P}^n$ be a standard determinantal scheme of codimension $c+1$. Thus there is a homogeneous $t \times (t+c)$ matrix $A$ such that the homogeneous ideal $I_V$ is generated by the maximal minors of $A$. If $t = 1$ then $V$ is a complete intersection and there is nothing to prove. Let $t > 1$. Then our assertion follows by induction if we have shown that $V$ is evenly G-linked to a standard determinantal scheme $V'$ whose homogeneous ideal is generated by the maximal minors of a $(t-1) \times (t+c-1)$ matrix $A'$. Actually, we will even prove that $A'$ can be chosen as the matrix which we get after deleting an appropriate row and column of the matrix $A$ and that then $V$ and $V'$ are directly G-linked in two steps. In order to do that we proceed in several steps.

*Step* I: Let $B$ be the matrix consisting of the first $t + c - 1$ columns of $A$. Then the ideal $I(B)$ has codimension $c$ according to [**9**]. Thus $I(B)$ defines a standard determinantal scheme $S$.

We claim that $S$ is even good determinantal. Indeed, the fact $I_V = I_t(A) \subset I_{t-1}(B)$ implies $\operatorname{codim} I_{t-1}(B) \geq \operatorname{codim} I_V = c + 1$. Thus the assertion follows by Proposition 3.2.

Since $S$ is good determinantal we may assume that the maximal minors of the matrix $A'$ consisting of the first $t - 1$ rows of $B$ generate an ideal which defines a standard determinantal scheme $V'$, i.e. $I_{V'} = I(A')$.

*Step* II: Let $A_1$ be the submatrix of $A$ consisting of its first $t - 1$ columns and put $J = I(A_1)$. Let $D$ be the determinant of the matrix which consists of the first $(t-1)$ and the last column of $A$. We claim that it holds

$$I_S : D = I_S \quad \text{and} \quad I_V = (I_S + (D)) : J.$$

In order to prove this we use techniques similar to those in [**45**]. Let $\varphi : F \to G$ and $\psi : F' \to G$ be the homomorphisms of free modules defined by $A$ and $B$, respectively. Since $B$ arises by deleting a column of $A$ we may assume that $F = F' \oplus R(b)$ for some integer $b$. Thus there is a commutative exact diagram

$$
\text{(3.1)}\qquad
\begin{array}{ccccccc}
0 & & & & & & \\
\downarrow & & & & & & \\
F' & \xrightarrow{\psi} & G & \to & M_\psi & \to & 0 \\
\downarrow & & \| & & & & \\
F & \xrightarrow{\varphi} & G & \to & M_\varphi & \to & 0 \\
\downarrow & & & & & & \\
R(b) & & & & & & \\
\downarrow & & & & & & \\
0. & & & & & &
\end{array}
$$

The Snake Lemma implies that there is a homogeneous ideal $I$ such that we get the following exact sequence

$$0 \to R/I(b) \xrightarrow{\varepsilon} M_\psi \to M_\varphi \to 0.$$

Due to Step I the ideals $I(\varphi)$ and $I(\psi)$ are standard determinantal. Therefore we know by [**12**] that

$$\operatorname{Ann} M_\varphi = I_V \quad \text{and} \quad \operatorname{Ann} M_\psi = I_S.$$

Thus considering the annihilators the exact sequence above implies $I_S \subset I$ and $I \cdot I_V \subset I_S$. Since $\operatorname{codim} I_V > \operatorname{codim} I_S$ the latter relation gives $I \subset I_S$. Altogether we conclude that $I = I_S$. Moreover Proposition 3.4(ii) shows that $\varepsilon$ maps $(1 \mod I_S)$ onto the residue class of $D$. Therefore the injectivity of $\varepsilon$ yields $I_S : D = I_S$.

Since $\operatorname{im} \varepsilon = (I_S + (D))/I_S$ we obtain by Proposition 3.4 that $(I_S + J)/(I_S + (D))$ is (up to a degree shift) isomorphic to $M_\varphi$, whose annihilator is $I_V$. It follows that

$$I_V = (I_S + (D)) : (I_S + J) = (I_S + (D)) : J,$$

and thus our claim is proved.

*Step* III: We will show that the ideal $I_S + D \cdot J^{c-1}$ is a Gorenstein ideal of codimension $c+1$ which is contained in $I_V$ and has degree

$$\deg(I_S + D \cdot J^{c-1}) = \deg D \cdot \deg S + \deg(I_S + J^{c-1})$$

where we define $J^0 = R$ and $\deg R = 0$.

Indeed, since $I_S + D \cdot J^{c-1}$ is generated by $t$-minors of $A$ the containment relation is clearly true.

Due to the previous step we know that $I_S : D = I_S$. Hence the multiplication by $D$ induces an isomorphism

$$[(I_S + J^{c-1})/I_S](-\deg D) \cong (I_S + D \cdot J^{c-1})/I_S.$$

Now we can conclude as in Corollary 3.5 that $I_S + D \cdot J^{c-1}$ is a Gorenstein ideal as well as is $I_S + J^{c-1}$.

Next, we consider the sequence

$$0 \to I_S(-\deg D) \to I_S \oplus (I_S + J^{c-1})(-\deg D) \to I_S + D \cdot J^{c-1} \to 0$$

where the first map is given by $R \mapsto (DR, R)$ (observe that $D \in J$) and the second map is given by $(A, B) \mapsto A - BD$. Using the first claim of the previous step again, it follows that this sequence is exact. Thus it gives the assertion on the degree. (This is similar to Lemma 4.8 on Basic Double G-Linkage.)

*Step* IV: We claim that for $i = 0, \dots, c$ it holds
$$\deg(I_S + J^i) = i \cdot [\deg D \cdot \deg S - \deg V].$$
In order to show this let us assume first that $V$ is good determinantal. Then we may also assume that $J + I_V$ has codimension $c + 2$.

By the previous step there is an exact sequence
$$0 \to R/I_S(b) \xrightarrow{\varepsilon} M_\psi \to M_\varphi \to 0.$$
Taking symmetric powers we obtain the exact sequence
$$S_i(M_\psi) \otimes R/I_S(b) \xrightarrow{\alpha} S_{i+1}(M_\psi) \to S_{i+1}(M_\varphi) \to 0.$$
Since $V$ is good determinantal Proposition 3.4 applies to $V$ and $S$. It follows (up to degree shift) that
$$\operatorname{im} \alpha \cong (I_S + D \cdot J^i)/I_S \cong [(I_S + J^i)/I_S](-d)$$
where we put $d = \deg D$. Thus the last sequence gives for $i = 0, \dots, c-1$ the exact sequence
$$0 \to [(I_S + J^i)/I_S](-d) \to (I_S + J^{i+1})/I_S \to (I_V + J^{i+1})/I_V \to 0.$$
Therefore we get for the Hilbert functions, using $I_S : D = I_S$:
$$\begin{aligned} &\operatorname{rank}_k[R/I_V]_j + \operatorname{rank}_k[R/(I_S + J^{i+1})]_j - \operatorname{rank}_k[R/(I_S + J^i)]_{j-d} \\ = \; &\operatorname{rank}_k[R/(I_V + J^{i+1})]_j + \operatorname{rank}_k[R/I_S]_j - \operatorname{rank}_k[R/I_S]_{j-d} \\ = \; &\operatorname{rank}_k[R/(I_V + J^{i+1})]_j + \operatorname{rank}_k[R/(I_S + (D))]_j. \end{aligned}$$
Since we know
$$\operatorname{codim} V = \operatorname{codim}(I_S + J^{i+1}) = \operatorname{codim}(I_S + (D)) < \operatorname{codim}(I_V + J^{i+1})$$
due to Corollary 3.5 we obtain
$$\deg(I_S + J^{i+1}) - \deg(I_S + J^i) = d \cdot \deg S - \deg V.$$
Now a simple induction shows the claim.

It remains to do the general case where $V$ is not necessarily good determinantal. For this we consider the exact sequence
$$0 \to S_i(M_\varphi)(-ia) \to R/I_S \to R/(I_S + J^i) \to 0, \quad 1 \le i \le c.$$
The first two modules in that sequence have a free resolution involving only symmetric and exterior powers of $F'$ and $G$, respectively. This means that their Hilbert functions depend only on the degrees of the entries of the matrix $A$. Similarly the Eagon-Northcott complex provides that the degree of $V$ depends only on the degrees of the entries of the matrix $A$. Now let us deform the entries of $A$ while keeping their degrees in order to get a matrix defining a good determinantal scheme. Since this deformation does not change the degrees of the ideals involved in our claim the asserted formula follows by the first part of the argument.

*Step* V: According to Step III we can G-link $I_V$ by $I_S + D \cdot J^{c-1}$. Let $I$ be the residual ideal. We claim that $I = I_S + J^c$.

In order to show this we note first that $I$ is a perfect ideal, too. Second, we use Step II. It provides
$$I_S + J^c \subset (I_S + D \cdot J^{c-1}) : I_V = I.$$
Indeed, due to Step II an element $P \in R$ belongs to $I_V$ if and only if $P \cdot J \subset I_S + DR$. It follows that $P \cdot J^c \subset I_S + D \cdot J^{c-1}$ as claimed.

Third we use Steps III and IV in order to compare the degrees of the two ideals:

$$\begin{aligned}\deg I &= \deg(I_S + D \cdot J^{c-1}) - \deg V \\ &= d \cdot \deg S + \deg(I_S + J^{c-1}) - \deg V \\ &= d \cdot \deg S + (c-1) \cdot [d \cdot \deg S - \deg V] - \deg V \\ &= c \cdot [d \cdot \deg S - \deg V] \\ &= \deg(I_S + J^c).\end{aligned}$$

Thus we know that $I$ and $I_S + J^c$ are perfect ideals of codimension $c+1$ (by Corollary 3.5), that they have the same degree and that $I_S + J^c \subset I$. Therefore both ideals must be equal as we wanted to show.

*Step* VI: Let $D'$ be the determinant of the matrix which consists of the first $(t-1)$ columns of $A'$. We want to sketch that we can repeat Steps II - V replacing $D$ by $D'$ and $V$ by $V'$.

Indeed, let $\varphi'$ be the homomorphism defined by $A'$. Since $A'$ arises by deleting a row of $B$ there is a free module $G'$ such that we have an exact commutative diagram as follows:

$$\begin{array}{ccccccc} & & 0 & & & & \\ & & \downarrow & & & & \\ & & R(b') & & & & \\ & & \downarrow & & & & \\ F' & \xrightarrow{\psi} & G & \rightarrow & M_\psi & \rightarrow & 0 \\ \| & & \downarrow & & & & \\ F' & \xrightarrow{\varphi'} & G' & \rightarrow & M_{\varphi'} & \rightarrow & 0 \\ & & \downarrow & & & & \\ & & 0. & & & & \end{array}$$

The Snake Lemma and consideration of the annihilators imply the exact sequence

$$0 \rightarrow R/I_S(b') \xrightarrow{\varepsilon'} M_\psi \rightarrow M_{\varphi'} \rightarrow 0$$

where $\varepsilon'$ maps $(1 \mod I_S)$ onto the residue class of $D'$. Thus we get as in Step II

$$I_S : D' = I_S \quad \text{and} \quad I_{V'} = (I_S + (D')) : J.$$

Furthermore, we see that the ideal $I_S + D' \cdot J^{c-1}$ is a Gorenstein ideal of codimension $c+1$ which is contained in $I_{V'}$ and has degree

$$\deg(I_S + D' \cdot J^{c-1}) = \deg D' \cdot \deg S + \deg(I_S + J^{c-1}).$$

Next, the arguments in Step IV provide

$$\deg(I_S + J^{i+1}) = i \cdot [\deg D' \cdot \deg S - \deg V'], \quad 0 \le i \le c-1.$$

Finally, we consider the ideal $I$ which is G-linked to $I_{V'}$ by $I_S + D'J^{c-1}$. It turns out that $I$ and $I_S + J^c$ are perfect of codimension $c+1$ and of the same degree. Since $I_S + J^c \subset I$ we must have equality.

*Step* VII: In the two last steps we have seen that it holds

$$(I_S + D \cdot J^{c-1}) : I_V = I_S + J^c = (I_S + D' \cdot J^{c-1}) : I_{V'}.$$

Since G-linkage is symmetric this means that $V$ is linked to $V'$ in two steps completing the proof. $\square$

REMARK 3.7. (i) Gaeta's original result says that every ACM subscheme of codimension two is licci. Since it is known for subschemes of codimension two that ACM subschemes are standard determinantal and that arithmetically Gorenstein subschemes are complete intersections, our theorem is a full generalization of Gaeta's Theorem.

(ii) Let $V \subset \mathbb{P}^n$ be a standard determinantal scheme defined by the maximal minors of a $t \times (c+t)$ matrix $A$. Then the proof above shows that $V$ can be G-linked in $2(t-1)$ steps to a complete intersection whereas the "standard" proof in codimension two ($c = 1$) shows that in this case $V$ can be linked in $t-1$ steps to a complete intersection. It is fun to compare the two proofs in case $c = 1$. The "standard" proof proceeds by showing that $V$ can be linked to the standard determinantal scheme $W$ which is defined by the maximal minors of the matrix which arises after deleting two columns of $A$. It turns out that $I_W$ is just the ideal $I_S + J$ using the notation of the previous proof. Thus our proof would also give the "standard" proof in case $c = 1$. However, we can not follow this simpler strategy if $c \geq 2$ since then the ideal $I_S + J^c$ will be no longer standard determinantal in general.

(iii) It is known that in general standard determinantal schemes of codimension $\geq 3$ are not licci. Consider for example a $t$-tuple point $P$. It is standard determinantal and has a linear minimal free resolution. Thus it is not licci according to [**36**], Corollary 5.13.

(iv) Gaeta's result mentioned above would be generalized even further by an affirmative answer to the following question: *Is every ACM subscheme glicci*? Even in codimension 3 this would be a very interesting result. See Chapter 8 for some progress in this direction.

CHAPTER 4

# Divisors on an ACM Subscheme of Projective Space

From now on, unless specified otherwise, $S$ will denote an ACM subscheme of $\mathbb{P}^n$ which satisfies property $G_1$ (see Definition 2.1). This will allow us to produce G-links in a rather straight-forward, geometric way. However, already in Chapter 3 we gave a very powerful result which avoided the use of this assumption (Theorem 3.6). In view of this, we will try to state our results avoiding the condition of satisfying property $G_1$ whenever possible, while at the same time trying to emphasize the geometric aspects of the constructions.

A *divisor* on $S$ will be an equidimensional, locally Cohen-Macaulay, codimension one subscheme of $S$. Since our main applications are for curves on surfaces, we will usually denote such a divisor by $C$. When we assume that $S$ satisfies $G_1$ (remembering that $S$ is always ACM), we can consider these as effective generalized divisors in the sense of Hartshorne [**30**] (see for instance his Proposition 2.4).

If $F$ is a homogeneous polynomial not vanishing on any component of $S$ (i.e. $I_S : F = I_S$), then $H_F$ is the codimension one subscheme of $S$ cut out by $F$. It is again ACM, and its homogeneous ideal (as a subscheme of $\mathbb{P}^n$) is given by $I_S + (F)$. We will sometimes view it as a subscheme of $\mathbb{P}^n$ and sometimes as a divisor on $S$.

We first give a simple example to illustrate the kinds of thing that we will have to be careful about when we do not have $G_1$.

EXAMPLE 4.1. Let $V \subset \mathbb{P}^3$ be the scheme defined by the square of the ideal $I_P = (x, y, z)$ supported at the point $P = [0, 0, 0, 1]$. Then $V$ is standard determinantal, defined by the maximal minors of the $2 \times 4$ matrix

$$\begin{bmatrix} x & y & 0 & z \\ 0 & y & z & x+y \end{bmatrix}.$$

$V$ is clearly not good determinantal, since removing a generalized row gives a $1 \times 4$ matrix whose ideal of maximal minors has height 3 rather than 4. Let $S$ be the ACM curve obtained by removing the last column of this matrix. Then $S$ consists of the union of three lines through the point $P$, with defining ideal $I_S = (xy, xz, yz)$. Unfortunately, $S$ does not satisfy $G_1$, although it does satisfy $G_0$! Yet Theorem 3.6 shows that $V$ is G-bilinked to $P$ (see also [**55**] Remark 5.1, where it is shown that in fact they are directly G-linked).

For comparison with the discussion below, notice that if $L$ is a linear form not vanishing on any component of $S$ then the corresponding hyperplane cuts out a subscheme $H_L$ on $S$ which is a zeroscheme of degree 3. However, if $L$ vanishes at $P$ then this zeroscheme is in fact defined by the square of the ideal of $P$ inside the hyperplane $H = \mathbb{P}^2$ defined by $L$. This shows that without $G_1$ we cannot always expect the effective divisor $H_L - P$ on $S$ to have the expected degree, since if $P$ is a point in $\mathbb{P}^2$ then $I_P^2 : I_P = I_P$. Compare this with Lemma 4.2.

LEMMA 4.2. *Let $C$ be a divisor on $S$, where $S$ is ACM satisfying $G_1$. Let $F \in I_C$ be a homogeneous polynomial of degree $d$ not vanishing on any component of $S$ (i.e. $I_S : F = I_S$), and let $H_F$ be the corresponding hypersurface section of $S$. Then the ideal $[I_S + (F)] : I_C$ gives a well-defined subscheme $C'$ of $S$. We will usually denote this by $C' = H_F - C$. The degree of $C'$ is $d \cdot \deg S - \deg C$. As subschemes of $\mathbb{P}^n$, $C$ and $C'$ are Cohen-Macaulay-linked by the ACM scheme $S \cap F$.*

PROOF. We have to show that $(I_S + (F)) : I_C = I_{C'}$ and $(I_S + (F)) : I_{C'} = I_C$. This clearly holds locally at any minimal prime ideal of $C$ or $C'$, since $S$ is assumed to satisfy $G_1$, and hence $S \cap F$ satisfies $G_0$. Since ideal quotients commute with localization, the result holds as stated. □

DEFINITION 4.3. Let $C$ be a divisor on $S$, where $S$ is ACM satisfying $G_1$. In the situation of Lemma 4.2, we say that *$C'$ is linked to $C$ on $S$ by the hypersurface divisor $H_F$*, and we denote this by $C \sim_F C'$ (when $S$ is understood). A divisor $C'$ is said to be *linearly equivalent* to $C$ on $S$ if there exists a divisor $Y$ and homogeneous polynomials $F \in I_C$ and $G \in I_{C'}$ of the same degree such that $Y = H_F - C = H_G - C'$. (Equivalently, $C \sim_F Y \sim_G C'$.) The linear system $|C|$ is the set of all such divisors.

Notice that this definition agrees with that of Hartshorne [**30**].

REMARK 4.4. (i) Example 4.1 shows that Lemma 4.2 does not hold without the assumption of $G_1$.
(ii) We would like to write "$C+Y = H_F$" in the above definition, but as Hartshorne [**30**] remarks, we have to be careful: if $C$ and $C'$ are divisors on $S$ then $C + C'$ does not always make sense.

Here are two examples. First, if $C = C'$ is a line in $\mathbb{P}^3$ and $S$ is the union of three planes passing through $C = C'$ then $S$ satisfies $G_1$ (in fact it is a complete intersection) but $C + C'$ does not make sense (at least in a degree-preserving way, as we would like).

Second (see also [**30**] Warning 4.1.2), if $S$ is the union of *two* planes passing through $C = C'$ as above, and $H$ is a general plane passing through $C = C'$, then $S \cap H$ algebraically links $C$ to $C'$. But as a subscheme of $S$, it does not make sense to write $C + C' = S \cap H$ since any other hyperplane through $C$ would also link $C$ to $C'$. However, it still makes sense to write $C' = (S \cap H) - C$, invoking linkage.
(iii) It is clear from Definition 4.3 that linear equivalence is preserved under hypersurface sections: if $C$ is linearly equivalent to $C'$ on $S$ and if $F$ is a general homogeneous polynomial then the schemes $C \cap F$ and $C' \cap F$ are linearly equivalent on $S \cap F$.

REMARK 4.5. Lemma 4.2 can be generalized as follows, to define the difference of divisors on a projective subscheme $S$ (with no further assumption on $S$). Let $C$ and $Y$ be divisors on $S$, with $I_Y \subset I_C$. Then $C' = Y - C$ makes sense as an effective divisor, with defining ideal $I_{C'} = I_Y : I_C$, if $Y$ satisfies property $G_0$. Notice that Lemma 4.2 is a special case since if $S$ is ACM and satisfies $G_1$ then $H_F$ satisfies $G_0$. Furthermore, if $S$ is ACM and satisfies property $G_1$ then we can also define $Y - C$ as an equivalence class of divisors, and it will be effective if and only if there is a divisor $D$, linearly equivalent to $C$, such that $I_Y \subset I_D$ and $Y$ satisfies $G_0$.

LEMMA 4.6. *Let $X$ be a divisor on $S$ which is linearly equivalent to a degree $d$ hypersurface section. Then $X$ actually is a hypersurface section: there exists some* *homogeneous polynomial $F$ of degree $d$ such that $X = H_F$. We denote this linear system by $|dH|$.*

PROOF. This follows from the fact that the restriction $H^0(\mathbb{P}^n, \mathcal{O}_{\mathbb{P}^n}(d)) \to H^0(S, \mathcal{O}_S(d))$ is surjective, since $S$ is ACM. (In fact, for this result it is enough to assume that $S$ is projectively normal.) More precisely, suppose $G$ is a polynomial of degree $d$ and suppose that there exists a divisor $Y$ and polynomials $G_1$ and $G_2$ of the same degree such that $X \sim_{G_1} Y \sim_{G_2} H_G$. Then $\frac{GG_1}{G_2}$ restricts to a global section of $\mathcal{O}_S(d)$ defining the divisor $X$, and hence by the surjectivity just mentioned, we have the desired result. □

LEMMA 4.7. *Let $C$ be a divisor on $S$. Let $C'$ be any effective residual divisor on $S$, i.e. $C' = H_F - C$ for some homogeneous polynomial $F \in I_C$ of degree $d$ not vanishing on any component of $S$. Then a divisor $Z$ is linearly equivalent to $C'$ if and only if there is a homogeneous polynomial $G \in I_C$ of degree $d$ such that $Z = H_G - C$.*

PROOF. The proof is very similar to the proof of Lemma 4.6. Assume that $C' \sim_{G_1} Y \sim_{G_2} Z$, where $G_1$ and $G_2$ have the same degree, and consider $\frac{FG_2}{G_1}$. □

We now describe algebraically the notion of "adding a hypersurface section," with almost no hypotheses, and then we will apply it to our situation.

LEMMA 4.8. *(**Basic Double G-Linkage**) Let $I \subset J$ be homogeneous ideals of $R = k[x_0, \dots, x_n]$ such that* codim $I + 1 =$ codim $J = c + 1$. *Let $A \in R$ be an element of degree $d$ such that $I : A = I$. Then it holds:*

(i) $\deg(I + A \cdot J) = d \deg I + \deg J$.
(ii) *If $I$ is perfect and $J$ is unmixed then $I + A \cdot J$ is unmixed.*
(iii) $J/I \cong (I + A \cdot J)/I(d)$.
(iv) *If $R/I$ and $J/I$ are Cohen-Macaulay and $J/I$ has Cohen-Macaulay type* 1 *then $J$ and $I + A \cdot J$ are Gorenstein ideals of codimension $c + 1$.*

PROOF. (sketch) Consider the sequence of $R$-modules

$$0 \to I(-d) \to I \oplus J(-d) \to I + A \cdot J \to 0 \tag{4.1}$$

where the first map is given by $F \mapsto (AF, F)$ and the second map is given by $(F, G) \mapsto F - AG$. Using $I : A = I$, it follows that this sequence is exact. This implies the first claim.

If $I$ is perfect then the exact sequence provides

$$H^i_{\mathfrak{m}}(J)(-d) \cong H^i_{\mathfrak{m}}(I + A \cdot J) \quad \text{if } i \leq n - c - 1.$$

Hence the cohomological unmixedness criterion [**57**], Lemma 2.12 shows that $J$ is unmixed if and only if $I + A \cdot J$ has this property.

In order to establish claim (iii) we consider the homomorphism induced by multiplication by $A$. Since $I : A = I$ it is an isomorphism.

Finally, we apply the Horseshoe Lemma ([**74**] 2.2.8, p. 37) to the exact sequence

$$0 \to I \to J \to J/I \to 0.$$

It shows that $J$ is perfect of Cohen-Macaulay type 1, thus it is a Gorenstein ideal. Using claim (iii), the same argument applied to the exact sequence

$$0 \to I \to I + A \cdot J \to J/I(-d) \to 0$$

yields that $I + A \cdot J$ is a Gorenstein ideal as well, which completes the proof. □

REMARK 4.9. Note that in Lemma 4.8 (iv) we do not claim that such an ideal $J$ necessarily exists. However, we will see in Lemma 5.2 that if $I$ defines an arithmetically Cohen-Macaulay subscheme $S$ possessing property $G_0$ then such an ideal $J$ must exist.

REMARK 4.10. Let $S$ be ACM, satisfying $G_1$, and let $C$ be a divisor on $S$. We have seen above that the additive notation for divisors does not always make sense. However, if $F$ is a homogeneous polynomial of degree $d$ not vanishing on any component of $S$ then we shall denote by $C + H_F$ the divisor on $S$ defined by the ideal $I_S + F \cdot I_C$. If we set $A = F$, $I = I_S$ and $J = I_C$ then the exact sequence (4.1) gives the Hilbert function formula

$$H(C + H_F, t) = H(H_F, t) + H(C, t - d).$$

The term "Basic Double G-Linkage" used in Lemma 4.8 will be justified in Proposition 5.10, where we will see that the schemes $C$ and $C + H_F$ are G-bilinked (even if we assume only $G_0$!). We will also see, for instance in Example 7.11 that this is not true in general for CI-liaison. However, notice that if $S$ is a complete intersection then this construction coincides with the "standard" Basic Double Linkage (cf. [**48**], [**7**], [**25**], [**57**], [**52**]), and in this case $C$ and $C + H_F$ are CI-bilinked.

Let $G \in I_C$ and let $Y = H_G - C$. Then $C + H_F$ can also be obtained as $C + H_F = H_{FG} - Y$. In particular, $C + H_F$ is obtained from $C$ by adding a hypersurface section of $S$, and it can be obtained from $C$ also by a sequence of two links on $S$ (in the sense of Definition 4.3).

Our next goal is to prove that under complete intersections, geometric linkage and algebraic linkage define the same equivalence relation on equidimensional, codimension $c$ subschemes of $\mathbb{P}^n$, generalizing a theorem of Rao for codimension two ([**63**] Theorem 1.7). We have to be a little bit careful about what we mean here. If $C$ is not a generic complete intersection then clearly it does not participate in any geometric link. However, suppose that we have two subschemes $C$ and $C'$ of $\mathbb{P}^n$ which are both generic complete intersections, and such that there is a sequence of *algebraic* links

$$C \sim C_1 \sim \cdots \sim C_t \sim C'$$

where some of the $C_i$ are not generic complete intersections. It is conceivable that there is *no* sequence of geometric links starting at $C$ and ending at $C'$.

We have two solutions to this dilemma. First, for arbitrary codimension we follow Rao [**63**] and avoid this situation by assuming that all the schemes in the sequence of algebraic links are generic complete intersections. Then specializing to codimension two, we make an argument similar in spirit to that of Schwartau [**67**] (who solved the problem for curves in $\mathbb{P}^3$) to show that the two equivalence relations are the same even if non generic complete intersections are allowed to occur in the sequence of links. In our case we show that it is true even if we allow our schemes to be non locally Cohen-Macaulay, assuming only that the corresponding ideals are unmixed. We will use the term "equidimensional" to include "no embedded components," i.e. that the corresponding ideal is unmixed.

Beginning with the case of arbitrary codimension, we make the following definition. Note that this differs from the corresponding definition of Rao [**63**] not only in allowing arbitrary codimension, but also in removing his assumption that the schemes be locally Cohen-Macaulay.

DEFINITION 4.11. Let $H(c,n)$ denote the set of all subschemes of $\mathbb{P}^n$ that are equidimensional of codimension $c$ and generic complete intersections.

LEMMA 4.12. *Let $C, C' \in H(c,n)$ be subschemes of $\mathbb{P}^n$ which are algebraically linked by a complete intersection, $X$, defined by $I_X = (F_1, \dots, F_c)$. Assume that deg $F_1 \leq \dots \leq$ deg $F_c$. After possibly performing a general change of basis, let $S$ be the complete intersection scheme defined by $I_S = (F_2, \dots, F_c)$. Then we can find a polynomial $G$ such that $I_S + (G) = (F_2, \dots, F_c, G)$ gives a* geometric *link of $C$ to a generic complete intersection divisor $Y$ on $S$, and also a polynomial $G'$ such that $I_S + (G') = (F_2, \dots, F_c, G')$ gives a* geometric *link of $C'$ to a generic complete intersection divisor $Y'$ on $S$. In fact, such a $G$ or $G'$ exist in any degree $\gg 0$.*

PROOF. If $C$ and $C'$ have no common component then there is nothing to prove, so we assume that there are common components. Let $\wp$ be a generic point at a common component. The proof will be similar in spirit to the proofs of Lemma 3 and Proposition 6 of [**65**].

We start with a minimal free resolution for $I_C$,

$$0 \to A_n \to \cdots \to A_c \to A_{c-1} \to \cdots \to A_2 \to \bigoplus_{i=1}^{m} R(-a_i) \to I_C \to 0.$$

Sheafifying, we obtain a kernel, $\mathcal{E}$, at the $c$-th step which is a reflexive sheaf. Suppose that deg $F_i = d_i$ and let $d = \sum d_i$. We have a commutative diagram

$$(4.2)\qquad \begin{array}{ccccccccccccc} 0 & \to & \mathcal{E} & \to & \mathcal{A}_{c-1} & \to & \cdots & \to & \mathcal{A}_2 & \xrightarrow{\varphi} & \bigoplus_{i=1}^{m} \mathcal{O}_{\mathbb{P}^n}(-a_i) & \to & \mathcal{I}_C & \to & 0 \\ & & \uparrow & & & & & & & & \uparrow \sigma & & \uparrow & & \\ 0 & \to & \mathcal{O}_{\mathbb{P}^n}(-d) & \to & \cdots & & & & & \longrightarrow & \bigoplus_{i=1}^{c} \mathcal{O}_{\mathbb{P}^n}(-d_i) & \to & \mathcal{I}_X & \to & 0. \end{array}$$

We know ([**58**] Corollary 1.5) that a mapping cone gives an exact sequence

$$(4.3)\quad 0 \to \bigoplus_{i=1}^{m} \mathcal{O}_{\mathbb{P}^n}(a_i - d) \xrightarrow{\begin{bmatrix} \varphi^\vee \\ -\sigma^\vee \end{bmatrix}} \mathcal{A}_2^\vee(-d) \oplus \bigoplus_{i=1}^{c} \mathcal{O}_{\mathbb{P}^n}(d_i - d) \to \cdots \to \mathcal{I}_{C'} \to 0.$$

Let $(4.2')$ and $(4.3')$ be the resolutions and diagrams obtained by localizing at $\wp$. Then both the top row of $(4.2')$ and $(4.3')$ are free resolutions of complete intersections. In particular, in $(4.2')$ we can split off to obtain $m = c$. But then since $(4.3')$ is a free resolution of the complete intersection $\mathcal{I}_{C',\wp}$, the Cohen-Macaulay type is one. Hence we can split off all but one of the summands of $\bigoplus_{i=1}^{c} \mathcal{O}_\wp(a_i - d)$; thus all but one of the generators, $F_i$, of $I_X$ have images in $\mathcal{I}_{C,\wp}$ which are minimal generators. Any homogeneous polynomial $G$ whose image in $\mathcal{I}_{C,\wp}$ provides the last generator of $\mathcal{I}_{C,\wp}$ will give the desired geometric link for $C$. Since a similar argument works for $C'$, a general choice of the generators of $I_X$ will work simultaneously for $C$ and $C'$, providing the desired complete intersection $S$. □

LEMMA 4.13. *Let $C \subset S$ be as in Lemma 4.12. Let $Y$ be a divisor which is linearly equivalent to $C$ on $S$, and let $Z$ be a basic double G-link divisor obtained from $C$, i.e. $Z = C + H_F$. Assume that $F$ does not vanish on any component of $C$. Then both $Y$ and $Z$ can be obtained from $C$ by a sequence of two geometric links. (The same holds for $C'$.)*

PROOF. Since $S$ is a complete intersection, the notion of divisors being linked on $S$ by hypersurface divisors (Definition 4.3) actually gives complete intersection links in $\mathbb{P}^n$. The result then follows from Lemma 4.12, Lemma 4.7 and Remark 4.10. □

THEOREM 4.14. *Algebraic and geometric CI-linkage generate the same equivalence relation on $H(c,n)$. That is, if $C, C' \in H(c,n)$ and there is a sequence of algebraic CI-links $C \sim C_1 \sim \cdots \sim C_t \sim C'$ with each $C_i \in H(c,n)$ then there is a sequence of geometric CI-links from $C$ to $C'$.*

PROOF. Without loss of generality we can let $C, C' \in H(c,n)$ and assume that $C$ and $C'$ are algebraically linked by a complete intersection $(F_1, F_2, \ldots, F_c)$. Perform a change of basis as in Lemma 4.12, and restrict to the complete intersection $S$ defined in that Lemma.

Let $d = \deg F_1$. Then $C' = dH - C$. Let $a > d$ such that there is a homogeneous polynomial of degree $a$ giving a geometric link of $C$ to a residual divisor $Z_1$. Let $Z_2$ be a basic double link obtained from $C'$ using a sufficiently general homogeneous polynomial of degree $a - d$. One can check using Lemma 4.7 that $Z_1$ is linearly equivalent to $Z_2$. Then applying Lemma 4.13 gives that $C'$ can be obtained from $C$ by a sequence of five geometric links. □

We now assume that our schemes have codimension two and are equidimensional, and we extend Theorem 4.14.

REMARK 4.15. Let $C$ be a generic complete intersection of codimension two. In **[63]** Remark 1.5, Rao gives criteria (using **[62]** Lemme 3.5) for an algebraic link to yield another generic complete intersection, and for a link to be geometric. That is, let $I_X = (F_1, F_2) \subseteq I_C$ be a complete intersection and suppose $I_X : I_C = I_{C'}$. Then Rao observes

(i) $C'$ is also a generic complete intersection if and only if at each generic point $\eta$ of $C$, either $F_1$ or $F_2$ yields a minimal generator of $\mathcal{I}_{C,\eta}$. If $F_1 \in H^0(\mathbb{P}^n, \mathcal{I}_C(\nu_1))$ is given arbitrarily, then for $\nu_2 \gg 0$ there is an open set in $H^0(\mathbb{P}^n, \mathcal{I}_C(\nu_2))$, elements $F_2$ of which satisfy this condition.
(ii) The link is geometric if and only if at each $\eta$, $F_1$ and $F_2$ give a minimal basis of $\mathcal{I}_{C,\eta}$. If $\nu_1, \nu_2 \gg 0$ then there is an open set in $(I_C)_{\nu_1} \times (I_C)_{\nu_2}$, elements of which satisfy this condition. In fact, it is always possible to find $F_1$ and $F_2$ forming part of a minimal basis for $I_C$ and giving a geometric link.

Note that while Rao assumed that his schemes are locally Cohen-Macaulay, they hold equally well in our more general context. Our Lemma 4.12 is similar in spirit but for arbitrary codimension.

THEOREM 4.16. *Let $C, C' \in H(2,n)$ and assume that there is a sequence of algebraic links $C \sim C_1 \sim \cdots \sim C_t \sim C'$. Then there is a sequence of geometric links from $C$ to $C'$.*

PROOF. In this proof we will refer to **[57]** for linkage results. However, we remark that in the locally Cohen-Macaulay case these results were originally due to Rao **[64]**, and in the non locally Cohen-Macaulay case similar results were obtained by Nollet **[58]**. The main point that we would like to make here is that all the links performed in **[57]** Lemma 6.1 and Proposition 6.2 are of the type described in Remark 4.15 (i), where one of the generators for the complete intersection is allowed to be of large degree. Hence without loss of generality we can assume that all such links preserve the property of being a generic complete intersection.

Performing a geometric link on $C$ if necessary, without loss of generality we may assume that $C$ and $C'$ are evenly linked. It follows ([**57**] Corollary 3.11) that $I_C$ and $I_{C'}$ have so-called $N$-type resolutions of the following form:

$$0 \to \bigoplus_{i=1}^{s} R(-a_i) \to N \to I_C \to 0$$

$$0 \to \bigoplus_{i=1}^{s} R(-b_i) \to N(h) \to I_{C'} \to 0$$

where $N$ is reflexive and $H^n_{\mathfrak{m}}(N) = 0$. (To arrange that the same $N$ is in the middle, it is enough to add suitable direct sums of line bundles trivially to the left and center terms of the resolutions.) At this point we do not care whether the links which produce these resolutions preserve the generic complete intersection property; we just require the existence of this kind of resolution.

Now we are in a position to apply [**57**] Lemma 6.1 and Proposition 6.2, together with the observation about the links at the beginning of the proof, to guarantee that $C$ and $C'$ are algebraically linked by a sequence of steps each of which is a generic complete intersection. The result then follows from Theorem 4.14 above. □

REMARK 4.17. Theorem 4.14 and Theorem 4.16 leave open the question of whether the equivalence relation on $H(c, n)$ would remain the same if we allow sequences of links of the sort described before Definition 4.11, in codimension greater than 2.

CHAPTER 5

# Gorenstein Ideals and Gorenstein Liaison

It is well-known (cf. for instance [**66**]) that Gorenstein liaison behaves very similarly to complete intersection liaison. Many of the tools that one can use are the same, and perhaps most importantly, the deficiency modules are still preserved up to duals and shifts. We refer to [**52**] and [**57**] for extensive background and references.

There are two important reasons why Gorenstein liaison has not been used nearly as extensively as complete intersection liaison. First, most of the results and applications to date have been in codimension two, where the Gorenstein ideals are precisely the complete intersections. The second reason is that it is often difficult to find a "good" Gorenstein ideal, apart from complete intersections, which contains a given scheme.

The purpose of this paper is to show how in many ways, Gorenstein liaison behaves better than complete intersection liaison. The fact that all "standard" determinantal ideals are in the Gorenstein linkage class of a complete intersection (Theorem 3.6) is already a strong indication in that direction. In this chapter we give a construction which allows one to find "good" Gorenstein ideals containing a given scheme, at least in a very natural setting, and we use this to demonstrate some important properties of Gorenstein liaison which are *not* shared by complete intersection liaison.

We would like to stress the naturalness of the ideas described here. Indeed, if one takes the special case where $S$ is a hypersurface, or more generally a complete intersection, then all of the constructions and results used here coincide with the analogous constructions and results of liaison theory as described in [**30**] or [**49**]. See also Exemple 2.4 of [**62**].

REMARK 5.1. If $S \subset \mathbb{P}^n$ is an ACM subscheme then its canonical module is Cohen-Macaulay of Cohen-Macaulay type 1 as an $R$-module.

LEMMA 5.2. *Let $S \subset \mathbb{P}^n$ be an ACM subscheme satisfying $G_0$. Then there is a homogeneous Gorenstein ideal $J$ of $R$ with codimension* $\operatorname{codim} S + 1$ *such that $J$ contains $I_S$ and $K_S \cong J/I_S(t)$ for some integer $t$. Furthermore, any ideal $J$ for which this isomorphism holds is arithmetically Gorenstein.*

PROOF. According to our assumptions $K_S$ is a torsion free $R/I_S$-module of rank 1 (cf., for example, [**10**], Proposition 3.3.18). Therefore $K_S$ can be identified with an ideal of $R/I_S$. Thus there is a homogeneous ideal $J$ of $R$ containing $I_S$ such that $K_S \cong J/I_S(t)$ for some integer $t$. Now we distinguish two cases.

If $K_S(-t) \neq R/I_S$ then $(R/I_S)/[K_S(-t)] \cong R/J$ is Gorenstein due to [**10**], Proposition 3.3.18. The claim follows.

If $K_S(-t) = R/I_S$ then we choose a homogeneous element $A \in R$ of degree, say $d$, such that $I_S : A = I_S$. Arguing as in the proof of Lemma 4.8(iii) we see that $(I_S + (A))/I_S(-d) \cong R/I_S$ is Cohen-Macaulay of Cohen-Macaulay type 1 (as an $R$-module). Hence Lemma 4.8(iv) shows that $I_S + (A)$ is a Gorenstein ideal as claimed.

For the last statement, note that without the assumption of at least $G_0$, such a $J$ may not even exist. The proof of the statement is essentially contained in the proof of Lemma 5.4 below. □

REMARK 5.3. Boij [**6**] has shown a partial converse to the last result. In fact using the notation of Lemma 5.2, his Theorem 3.3 implies that for all Gorenstein ideals $J$ containing $I_S$ of sufficiently large initial degree there is an isomorphism $K_S \cong J/I_S(t)$ for some integer $t$.

We now give a geometric version of Lemma 5.2, although in order to use the language of generalized divisors we have to assume $G_1$.

LEMMA 5.4. *Let $S$ be an ACM subscheme of $\mathbb{P}^n$ satisfying property $G_1$, and let $X$ be a twisted canonical divisor on $S$ (i.e. a subscheme of $S$ defined by the vanishing of a regular section of $\omega_S(\ell)$ for some $\ell \in \mathbb{Z}$). Let $F \in I_X$ be a homogeneous polynomial of degree $d$ such that $F$ does not vanish on any component of $S$. Let $H_F$ be the divisor cut out on $S$ by $F$. Then the (effective) divisor $H_F - X$ on $S$, viewed as a subscheme of $\mathbb{P}^n$, is arithmetically Gorenstein. We will call such a divisor a "twisted anticanonical divisor." In fact, any effective divisor in the linear system $|H_F - X|$ is arithmetically Gorenstein.*

PROOF. We are assuming that $X$ is the divisor associated to a regular section of $\omega_S(\ell)$. Let $Y$ be the residual divisor, $Y \in |H_F - X| = |dH - X|$. Let $I_Y$ be the homogeneous ideal of $Y$, viewed as a subscheme of $\mathbb{P}^n$. Consider the ideal $I_Y/I_S$ in $A = R/I_S$ and its sheafification $\mathcal{I}_{Y|S}$ in $\mathcal{O}_S$. We have

$$\mathcal{I}_{Y|S}(d) \cong \mathcal{O}_S(dH - Y) \cong \mathcal{O}_S(X) \cong \omega_S(\ell).$$

Suppose that $S$ has codimension $r$. Since $S$ is ACM, the dual of the minimal free resolution of $I_S$ is a minimal free resolution for $H^0_*(\omega_S)$, and the last free module in this resolution has rank one. We thus have the following exact diagram (up to twist– what is important is the ranks):

$$\begin{array}{ccccccccc}
 & & & & & & 0 & & \\
 & & & & & & \downarrow & & \\
 & & 0 & & & & R & & \\
 & & \downarrow & & & & \downarrow & & \\
 & & F_r & & & & F_1^* & & \\
 & & \downarrow & & & & \downarrow & & \\
 & & F_{r-1} & & & & F_2^* & & \\
 & & \downarrow & & & & \downarrow & & \\
 & & \vdots & & & & \vdots & & \\
 & & \downarrow & & & & \downarrow & & \\
 & & F_1 & & & & F_r^* & & \\
 & & \downarrow & & & & \downarrow & & \\
0 & \rightarrow & I_S & \rightarrow & I_Y & \rightarrow & H^0_*(\omega_S)(\ell - d) & \rightarrow & 0. \\
 & & \downarrow & & & & \downarrow & & \\
 & & 0 & & & & 0 & &
\end{array}$$

The Horseshoe Lemma then shows that $Y$ is arithmetically Gorenstein. Note that this shows that any element in the linear system of $Y$ is arithmetically Gorenstein. □

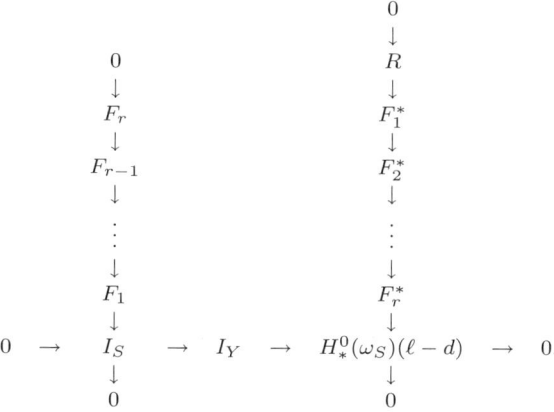

COROLLARY 5.5. *Let* $Y \subset S$ *be any effective divisor in the linear system* $|H_F - X|$ *as in Lemma 5.4. Let* $G$ *be a homogeneous polynomial not vanishing on any component of* $S$. *Then* $Y + H_G$ *is also arithmetically Gorenstein.*

PROOF. By Remark 4.10, $Y + H_G = H_{FG} - X$. □

REMARK 5.6. If $K$ is a canonical divisor on $S$, Lemma 5.4 and Corollary 5.5 amount to saying that *any* element of the linear system $|dH - K|$ is arithmetically Gorenstein. The fact that this holds for every element of the linear system, and not just on an open set, is what makes our Gorenstein liaison (described below) work.

EXAMPLE 5.7. Suppose $X \subset \mathbb{P}^3$ is a good determinantal zeroscheme (cf. Definition 3.1), defined by the vanishing of the maximal minors of a homogeneous $t \times (t+2)$ matrix $A$ (if $t = 1$, $X$ is a complete intersection). Then it follows from [**45**] that there is a suitable local complete intersection, ACM curve $S$, containing $X$, such that $X$ is a twisted canonical divisor on $S$, and hence Lemma 5.4 applies: the residual to $X$ in any hypersurface section of $S$ containing $X$ is arithmetically Gorenstein.

EXAMPLE 5.8. It is not hard to show that if $\mathcal{D}$ is a linear system of divisors on $S$, then the graded Betti numbers of elements of $\mathcal{D}$ are upper semicontinuous. The fact that *any* element of the linear system of Lemma 5.4 turns out to be arithmetically Gorenstein is thus very surprising. Notice that the Lemma does not say anything about the other graded Betti numbers.

We give two examples, both on the Castelnuovo surface $S$ in $\mathbb{P}^4$. $S$ has degree 5 and sectional genus 2. The canonical divisor on $S$ has degree $-3$; in fact, $-K$ is a plane cubic.

For the first example, let $L$ be a general linear form. Then the curve $Y = -K + H_L$ has degree 8 and arithmetic genus 5. Its resolution has the form

$$0 \to R(-6) \to R(-4)^{\oplus 3} \oplus R(-3)^{\oplus 2} \to R(-2)^{\oplus 3} \oplus R(-3)^{\oplus 2} \to I_Y \to 0.$$

Notice that $Y$ could not possibly be the complete intersection of three quadrics since it has a plane cubic as a component. However, a general element of the linear system determined by $Y$ *is* the complete intersection of three quadrics (this can be checked, for example, with the computer program Macaulay [**5**]). Hence the graded Betti numbers (except the last one) change when we specialize to $Y$.

For the second example, again start with the plane cubic $-K$ but this time just add two exceptional divisors which are lines, say $Y = -K + E_2 + E_3$. This is a curve of degree 5, and its hyperplane section consists of five points, three of which are on a line. The main result of [**17**] shows that $Y$ cannot be arithmetically Gorenstein since its hyperplane section does not have the Cayley-Bacharach property. In fact, its minimal free resolution has the form

$$0 \to R(-4)^{\oplus 2} \oplus R(-5) \to R(-3)^{\oplus 6} \oplus R(-4)^{\oplus 2} \to R(-2)^{\oplus 5} \oplus R(-3) \to I_Y \to 0.$$

However, one can check that a general element of the linear system $|Y|$ *is* arithmetically Gorenstein– everything that “should” split off in the above resolution in fact does.

REMARK 5.9. We saw in Corollary 5.5 that if $C \subset S$ is arithmetically Gorenstein *of the type described in Lemma 5.4*, then adding a hypersurface section (i.e. performing a basic double G-link– cf. Lemma 4.8 and Remark 4.10) gives a curve which is again arithmetically Gorenstein. On the other hand, from the exact sequence

$$0 \to I_S(-d) \to I_C(-d) \oplus I_S \to F \cdot I_C + I_S \to 0,$$

a mapping cone construction for a free resolution of $F \cdot I_C + I_S$ would lead one to doubt very much that so many terms necessarily would split off to arrive at a Cohen-Macaulay type 1. The particular form of $C$ is important here. For example, if $S$ is the Castelnuovo surface in $\mathbb{P}^4$ and $C$ is a line on $S$, then $C$ is arithmetically Gorenstein. However, adding a hyperplane section gives a reduced curve $Y$ of degree 6 and arithmetic genus 2. Its hyperplane section consists of 6 points in $\mathbb{P}^3$, 5 of which lie on a plane. Hence the hyperplane section does not have the Cayley-Bacharach property, and so $Y$ cannot be arithmetically Gorenstein (again using the main result of [**17**]).

As promised in Remark 4.10, we now justify the term "Basic Double G-Linkage" assuming only $G_0$ for $S$. The method of proof generalizes the one used in order to establish Gaeta's Theorem.

PROPOSITION 5.10. *Let $S \subset \mathbb{P}^n$ be an ACM subscheme having property $G_0$. Let $C \subset S$ be an equidimensional subscheme of codimension* 1 *and let $A \in R$ be a homogeneous element such that $I_S : A = I_S$. Then $I_C$ and $I_S + A \cdot I_C$ are G-bilinked.*

PROOF. We will explicitly describe two Gorenstein ideals which will give the G-bilinkage. We proceed in several steps.

*Step* I: Let $c$ denote the codimension of $S$ and let $d$ denote the degree of $A$. According to Lemma 5.2 there is a Gorenstein ideal $J$ in $R$ of codimension $c+1$ such that $J/I_S$ is Cohen-Macaulay of Cohen-Macaulay type 1.

Since $\operatorname{codim} C > \operatorname{codim} S$ there is a homogeneous element $B \in I_C$ of degree, say $e$, such that $I_S : B = I_S$.

*Step* II: Due to Lemma 4.8 the ideal $I_S + B \cdot J$ is a Gorenstein ideal of codimension $c+1$. Hence the ideal $\mathfrak{a} := (I_S + B \cdot J) : I_C$ is (directly) G-linked to $I_C$. Thus $\mathfrak{a}$ is an unmixed ideal. Moreover, it follows by Lemma 4.8 that

$$\begin{aligned} \deg \mathfrak{a} &= \deg(I_S + B \cdot J) - \deg C \\ &= e \cdot \deg S + \deg(I_S + J) - \deg C. \end{aligned}$$

*Step* III: Our assumptions on $A$ and $B$ imply $I_S : AB = I_S$. Furthermore $I_S + A \cdot I_C$ is an unmixed ideal by Lemma 4.8. Therefore $I_S + AB \cdot J$ is a Gorenstein ideal and the ideal $\mathfrak{b} := (I_S + AB \cdot J) : (I_S + A \cdot I_C)$ is (directly) G-linked to $I_S + A \cdot I_C$ and in particular unmixed, too. Using Lemma 4.8 again and Step II we obtain

$$\begin{aligned} \deg \mathfrak{b} &= \deg(I_S + AB \cdot J) - \deg(I_S + A \cdot I_C) \\ &= (d+e) \cdot \deg S + \deg(I_S + J) - [d \cdot \deg S + \deg C] \\ &= \deg \mathfrak{a}. \end{aligned}$$

*Step* IV: Since we have by the definition of $\mathfrak{a}$ that $\mathfrak{a} \cdot I_C \subset I_S + B \cdot J$ we get $\mathfrak{a} \subset \mathfrak{b}$. But we have seen that $\mathfrak{a}$ and $\mathfrak{b}$ are unmixed ideals of the same degree and codimension. Hence they must be equal. Now the claim follows from Steps II and III. □

REMARK 5.11. (i) The last result generalizes previous notions of basic double linkage where $S$ was assumed to be (at least) arithmetically Gorenstein (cf. [**48**], [**25**], [**8**] and [**57**], Proposition 4.4). Taken in conjunction with Proposition 5.12 and Corollary 5.14 below, this also generalizes the notion of *elementary biliaison* of [**49**] and [**30**], where $S$ is assumed to be a complete intersection but linear equivalence is allowed.

(ii) If one follows the proof of Proposition 5.10 closely, one will note that the same proof gives that if $S$ is a zeroscheme with property $G_0$ (e.g. a reduced zeroscheme) and if $I$ is an artinian ideal containing $I_S$, then we can again form the ideal $I_S + A \cdot I$ and this will be G-bilinked to $I$.

We now turn to the fact that linearly equivalent divisors on an ACM subscheme $S$ of $\mathbb{P}^n$ are G-bilinked. We first give an algebraic formulation. Then we will give a geometric description which will be useful in the remaining chapters of this paper.

PROPOSITION 5.12. *Let $S \subset \mathbb{P}^n$ be an ACM subscheme of codimension $c$ having property $G_1$. Let $C, Y$ be generalized divisors on $S$ such that there are homogeneous polynomials $F \in I_C$ and $G \in I_Y$ with $I_S : F = I_S$, $I_S : G = I_S$ and $(I_S + (F)) : I_C = (I_S + (G)) : I_Y$. Then $C$ and $Y$ are G-bilinked.*

PROOF. We proceed in several steps.

*Step* I: Let $D$ be the divisor on $S$ defined by

$$I_D = (I_S + (F)) : I_C = (I_S + (G)) : I_Y.$$

Since $S$ has the property $G_1$ we obtain

$$\begin{aligned} \deg D &= \deg F \cdot \deg S - \deg C \\ &= \deg G \cdot \deg S - \deg Y. \end{aligned}$$

*Step* II: Consider a minimal free resolution of $I_S$:

$$0 \to \mathbb{F}_c \xrightarrow{A} \mathbb{F}_{c-1} \to \cdots \to I_S \to 0.$$

Notice that the transpose $A^t$ of $A$ is a presentation matrix for $K_S(n+1)$. Suppose that $A^t$ is a $t \times (t+r)$ matrix. Let $C$ be a sufficiently general $t \times 1$ matrix such that the concatenation, $A'$, of $A^t$ and $C$ is again homogeneous. Let $I$ be the ideal of the annihilator of the module $M_{A'}$ where $M_{A'}$ is the cokernel of the map represented by $A'$. We claim that the scheme $K$ defined by $I$ is a twisted canonical divisor of $S$.

Indeed, there is a commutative exact diagram

$$\begin{array}{ccccccc} 0 & & & & & & \\ \downarrow & & & & & & \\ \mathbb{F}_{c-1}^* & \xrightarrow{A^t} & \mathbb{F}_c^* & \to & K_S(n+1) & \to & 0 \\ \downarrow & & \| & & & & \\ \mathbb{F}_{c-1}^* \oplus R(b) & \xrightarrow{A'} & \mathbb{F}_c^* & \to & M_{A'} & \to & 0 \\ \downarrow & & & & & & \\ R(b) & & & & & & \\ \downarrow & & & & & & \\ 0. & & & & & & \end{array} \tag{5.1}$$

The Snake Lemma implies that there is a homogeneous ideal $I'$ such that we get the following exact sequence

$$0 \to R/I'(b) \to K_S(n+1) \to M_{A'} \to 0.$$

Thus considering the annihilators the exact sequence above implies $I_S = \operatorname{Ann} K_S \subset I'$ and $I' \cdot \operatorname{Ann} M_{A'} = I' \cdot I \subset I_S$. Since $Rad(I) = Rad(I_t(A'))$ and thus $\operatorname{codim} I' = \operatorname{codim} I_t(A') > \operatorname{codim} I_S$ the latter relation provides $I' \subset I_S$. Altogether we conclude that $I' = I_S$. Now our claim follows.

*Step* III: Since the submaximal minors of $A$ define the non locally Gorenstein locus of $S$, the ideal of submaximal minors of $A$ defines a subscheme $B$ of $S$ of codimension $\geq 2$ in $S$, since $S$ has property $G_1$. Let $P$ be any point of $S$ away from $B$. We see immediately that there is some submaximal minor of $A$ which does not vanish at $P$. It follows that for a sufficiently general choice of $C$ in Step II, the resulting scheme $K$ does not contain $P$. We conclude that we can choose $C$ so that the corresponding twisted canonical divisor $K$ has no component in common with $D$ (from Step I).

*Step* IV: Let $I_K$ be the saturation of $I$ (from Step II). Let $A \in I_K$ be sufficiently general, so that $I_S : A = I_S$ and $I_D : A = I_D$ (this last condition is valid by Step III). By Lemma 5.4, the ideal $J = [I_S + (A)] : I_K$ is Gorenstein. As in the proof of Lemma 5.4, we have that $J/I_S$ is Cohen-Macaulay with Cohen-Macaulay type 1. Since $I_D : A = I_D$, we get that the scheme defined by $J$ has no component in common with $D$. It follows that $\deg[I_D \cap J] = \deg D + \deg J$.

*Step* V: Let $F$ and $G$ be as in the statement of the proposition, and consider the ideals $I_S + F \cdot J$ and $I_S + G \cdot J$. By Lemma 4.8, both of these ideals define subschemes of $S$ which are Gorenstein of codimension 1 in $S$ (in particular unmixed), and which have degrees

$$\begin{aligned}\deg I_S + F \cdot J &= \deg F \cdot \deg S + \deg J \\ \deg I_S + G \cdot J &= \deg G \cdot \deg S + \deg J\end{aligned}$$

*Step* VI: Since $I_S + F \cdot J \subset I_C$, we see that the ideal $\mathfrak{a} = [I_S + F \cdot J] : I_C$ is (directly) G-linked to $I_C$. Hence

$$\begin{aligned}\deg \mathfrak{a} &= \deg F \cdot \deg S + \deg J - \deg C \\ &= \deg D + \deg J \\ &= \deg[I_D \cap J].\end{aligned}$$

In a similar way, the ideal $\mathfrak{b} = [I_S + G \cdot J] : I_Y$ is directly G-linked to $I_Y$ and has degree $\deg[I_D \cap J]$.

*Step* VII: By the definition of $\mathfrak{a}$ we know $\mathfrak{a} \cdot I_C \subset I_S + F \cdot J \subset I_S + (F)$. This implies $\mathfrak{a} \subset I_D$. Now let $T$ be an element of $\mathfrak{a}$. Since $F$ is in $I_C$ we obtain from $T \cdot I_C \subset I_S + F \cdot J$ in particular that $TF \in I_S + F \cdot J$. It follows that

$$T + J \subset [I_S + F \cdot J] : F \subset J,$$

and thus $T \in J$. Altogether we have shown that $\mathfrak{a} \subset I_D \cap J$. But both ideals are unmixed and have the same codimension and degree. Hence they must be equal.

*Step* VIII: Arguing as in the previous step we see that the ideal $\mathfrak{b}$ is (directly) G-linked to $I_Y$, is contained in $I_D \cap J$ and has the same degree as $I_D \cap J$. We conclude that $\mathfrak{b} = I_D \cap J = \mathfrak{a}$. Therefore, we have seen that $I_C$ is G-linked to $I_D \cap J$ which in turn is G-linked to $I_Y$. This means that $C$ and $Y$ are G-bilinked. □

EXAMPLE 5.13. Lemma 5.2 and Proposition 5.10 are very important in our theory, and one notices that they assume that $S$ is only $G_0$. On the other hand, Proposition 5.12 is central to much of what is to come, and it requires that $S$ be $G_1$. It is natural to ask if Proposition 5.12 might still be true with the weaker assumption of $G_0$. We now give a counterexample to show this is not the case. This example was first run on Macaulay [**5**].

Let $Y \subset \mathbb{P}^4$ be a line and let $S$ be the union of three general planes containing $Y$. For instance, in the ring $R = k[x_0, x_1, x_2, x_3, x_4]$ take $I_S = (x_0x_1, x_1x_2, x_0x_2)$ and $I_Y = (x_0, x_1, x_2)$. Choose a non-degenerate double line, $C$, of arithmetic genus $\leq -1$ supported on $Y$. $C$ is not arithmetically Cohen-Macaulay. The ideal $I_C$ can be obtained by adjoining two sufficiently general forms to the ideal $I_Y^2$ (and the resulting ideal is saturated). Choose a general element $F$ of degree 2 in $I_C$, and in Proposition 5.12 set $G := F \in I_Y \subset I_C$. Then one can check that $(I_S + (F)) : I_C = (I_S + (G)) : I_Y$. But $Y$ is arithmetically Cohen-Macaulay while $C$ is not, so they cannot be G-bilinked.

We now take a more geometric approach. *For the rest of this chapter, unless explicitly stated otherwise, we will consider schemes $C$ which are effective divisors on an ACM scheme $S$ having property $G_1$, and we will perform Gorenstein links on $S$ using Lemma 5.4 and Corollary 5.5.* Given $C$, our interest is to see which other divisors on $S$ are in the same Gorenstein liaison class as $C$, and we will study the linear system $|dH - K - C|$, where $K$ is a canonical divisor.

In a sense, our construction generalizes Hartshorne's description of liaison in [**30**]. There, he considers closed subschemes $V_1$ and $V_2$ of $\mathbb{P}^n$ which are both equidimensional of dimension $r$ (hence without embedded components, by our convention above). He says that "$V_1$ is linked to $V_2$ by a complete intersection $X$, if there is a complete intersection scheme $S$ of dimension $r+1$ containing $V_1$ and $V_2$, and a hypersurface $T$ containing $V_1$ and $V_2$, such that $S \cap T = X$, and $V_2 = X - V_1$ as generalized divisors on $S$." In other words, starting with $V_1$ on $S$ (and given $V_1$, such an $S$ can always be found), we perform a link on $S$ by choosing any hypersurface $T$ and looking at the residual divisor $H_T - V_1$. But any such divisor $H_T$ is in fact a twisted anticanonical divisor in the sense of Lemma 5.4, since $S$ is a complete intersection! Hence this complete intersection link is a special case of our Gorenstein links.

We now give the geometric version of Proposition 5.12. In the special case where $S$ is smooth we also give a much simpler proof; the problem with this proof in the more general case of $G_1$ is that it requires a suitable addition of divisors, which is avoided by Proposition 5.12. It should be compared also with Lemma 4.13. It shows a very nice property of Gorenstein linkage which is *not* shared by complete intersection linkage in codimension $\geq 3$ in $\mathbb{P}^n$ (see for instance Example 7.11) except in the special case of divisors on a complete intersection, as described in Lemma 4.13.

COROLLARY 5.14. *Let $C \subset S$ be a divisor, where $S$ is an ACM subscheme of $\mathbb{P}^n$ satisfying $G_1$. Let $Y$ be a divisor in the linear system $|C + tH|$, where $H$ is the hyperplane section class and $t \in \mathbb{Z}$. (Notice that if $t = 0$ then $C$ and $Y$ are linearly equivalent.) Then $Y$ can be obtained from $C$ by a sequence of two Gorenstein links.*

PROOF. This follows immediately from what was proved in Proposition 5.12 (take $\deg F + t = \deg G$).

We now assume that $S$ is smooth, and we give a simpler proof in the language of divisors. Throughout this proof, equality means "as divisors," not as linear equivalence classes. Let $K$ be an effective twisted canonical divisor on $S$. For $a \gg 0$ we can find a form $A$ of degree $a$ such that $H_A - K$ is effective; it is enough to choose $A \in I_K$ not vanishing on any component of $S$. We can also find forms $F \in I_C$ and $G \in I_Y$ with $\deg F + t = \deg G$, and a divisor $D$ on $S$, such that $H_F - C = D$ and $H_G - Y = D$. (This is the geometric equivalent of the assumption on the ideal quotients in the statement of Proposition 5.12. See also Lemma 4.7.)

Now we link. Observe that $H_{AF} - K$ is Gorenstein, by Remark 5.6. On the other hand, $H_{AF} - K - C = (H_A - K) + (H_F - C) = (H_A - K) + D$. The fact that this makes sense as a divisor, and is what we "expect" it to be (see Remark 4.4) follows from the fact that $H_{AF} - K$ is Gorenstein, so the residual is well-defined and respects degrees. In a similar way, $H_{AG} - K$ is Gorenstein and links $Y$ also to $(H_A - K) + D$. Notice that both $C$ and $Y$ are linked to the union of $D$ and a Gorenstein "tail." □

COROLLARY 5.15. *Let $S$ be an ACM surface satisfying $G_1$. Then the property that a divisor on $S$ is ACM, and the property that a divisor on $S$ is arithmetically Buchsbaum, are preserved under linear equivalence and under adding hypersurface sections.*

PROOF. This follows immediately from Corollary 5.14 and the fact that these properties are invariant under G-liaison ([**52**] Remark 5.3.2). □

REMARK 5.16. (i) Notice that Proposition 5.10 gives a special case of Corollary 5.14 under the weaker assumption that $S$ satisfies only $G_0$. That is, we can no longer consider linear equivalence, but we have the fact that the divisor $C + H_A$ is G-linked to $C$ in two steps.
(ii) In view of Corollary 5.14, we have that if $C$ and $C'$ are linearly equivalent on $S$ then their deficiency modules are isomorphic: $H^i_*(\mathbb{P}^n, \mathcal{I}_C) \cong H^i_*(\mathbb{P}^n, \mathcal{I}_{C'})$ for $1 \leq i \leq \dim C$. This holds even though $C$ and $C'$ are not necessarily in the same complete intersection liaison class (see for instance Example 7.11).

EXAMPLE 5.17. We give a simple example to show that Corollary 5.14, Corollary 5.15 and Remark 5.16 (ii) are false without the assumption that $S$ is ACM. Let $V$ be the Veronese surface in $\mathbb{P}^5$. Then $V$ is ACM of degree 4. Let $\Lambda = \mathbb{P}^4$ be a general hyperplane in $\mathbb{P}^5$ and let $S$ be the projection, from a general point in $\mathbb{P}^5$, of $V$ down to $\Lambda$. Then $S$ is a smooth, non ACM surface of degree 4 with $h^1(\mathbb{P}^4, \mathcal{I}_S(1)) = 1$. In particular, the sublinear system of $|\mathcal{O}_S(1)|$ cut out by hyperplanes is not complete, and a hyperplane section is not an ACM curve ([**52**] Theorem 1.3.2).

Let $C$ be the curve cut out on $V$ by $\Lambda$. Then $C$ is ACM, and it also lies on $S$. It is an element of $|\mathcal{O}_S(1)|$, so we have that the general element is ACM, while a special element is not. In particular, they are not G-linked.

EXAMPLE 5.18. The fact given in Remark 5.16 (ii) allows us to give a simple example to illustrate that our intuition, about which curves are linearly equivalent on a surface, is not always correct.

Let $S \subset \mathbb{P}^3$ be a surface consisting of the union of two planes, $\Lambda_1$ and $\Lambda_2$. Let $\lambda$ be the line of intersection of $\Lambda_1$ and $\Lambda_2$. Let $C = Y \cup R$ consist of the disjoint union of a conic, $Y$, on $\Lambda_1$ and a line, $R$, on $\Lambda_2$. Let $P$ be the point of intersection of the two lines $\lambda$ and $R$. Let $C' = Y' \cup R'$ be another such curve on $S$, and let $P'$ be the corresponding point. Then the deficiency module of $C$ is isomorphic to that of $C'$ *if and only if* $P = P'$ (cf. [**51**] Example 2.3). So even though $Y$ and $Y'$ move in a linear system on $\Lambda_1$ and $R$ and $R'$ move in a linear system on $\Lambda_2$, the same is not true for $C$ and $C'$ on $S$. See also [**30**], especially Chapter 5, where this example is studied extensively.

CHAPTER 6

# CI-Liaison Invariants

In this chapter we consider groups which are invariant under (algebraic) CI-linkage. Recall that two equidimensional, locally Cohen-Macaulay schemes $X \subset \mathbb{P}^N$ and $X' \subset \mathbb{P}^N$ are said to be in the same CI-liaison class if there exists a sequence $X = X_0, X_1, ..., X_t = X'$ of closed subschemes in $\mathbb{P}^N$ such that $X_i$ and $X_{i+1}$ are algebraically linked by some complete intersection $Y_i \subset \mathbb{P}^N$. A graded $R$-module $C(X)$ which depends only on $X$, is liaison-invariant as an $R$-module (resp. $k$-module) provided there exists a homogeneous $R$ (resp. $k$)-module isomorphism $C(X) \cong C(X')$ for any $X'$ in the CI-liaison class of $X$. A group $C(X \subset Y)$ (resp. a morphism of groups $C(X \subset Y) \to D(X \subset Y)$) which depends on $Y$ is "liaison-invariant" (here " " is usually used) provided $C(X \subset Y) \cong C(X' \subset Y)$ (resp. such isomorphisms exist and fit into the obvious commutative diagram) where $X \subset Y$ and $X' \subset Y$ are directly linked.

It is well known that the cohomology groups $H^i_*(\mathcal{I}_X)$, $1 \leq i \leq \dim X$, are liaison-invariant up to shift and dual (and in fact G-liaison-invariant by [**66**]), and that they all vanish if $X$ is ACM. (See [**52**] for a general reference.) In what follows we will consider non-trivial liaison-invariants of ACM schemes. They are in general not G-liaison-invariants (Remark 6.15). Indeed we will first use a rather natural geometric approach to give a new proof of an unpublished result of Buchweitz and Ulrich; the liaison-invariance of $H^i_{\mathfrak{m}}(K_{R/I_X} \otimes_R I_X)$ and of most $H^i_*(\mathcal{N}_X)$, where $K_{R/I_X}$ is the canonical module $\operatorname{Ext}^c_R(R/I_X, R)(-N-1)$ and $X = \operatorname{Proj}(R/I_X)$. To understand our geometric approach (originated in [**37**], cf. [**38**], Section 2), we need to recall some facts. If $X$ is equidimensional and locally Cohen-Macaulay, and if $X$ is a closed subscheme of a local complete intersection $Y \subset \mathbb{P} = \mathbb{P}^N$, i.e. $X \subset Y$ (with ideal sheaf $\mathcal{I}_{X|Y}$), we define the sheaf $A^1_{X|Y}$ by the *cartesian* diagram

$$(6.1) \qquad \begin{array}{ccccc} A^1_{X|Y} & \to & \mathcal{N}_Y & = & \mathcal{H}om_{\mathcal{O}_{\mathbb{P}}}(\mathcal{I}_Y, \mathcal{O}_Y) \\ \downarrow & \Box & \downarrow & & \\ \mathcal{N}_X & \to & \mathcal{N}_Y \otimes \mathcal{O}_X & \cong & \mathcal{H}om_{\mathcal{O}_{\mathbb{P}}}(\mathcal{I}_Y, \mathcal{O}_X). \end{array}$$

This and the exact sequence $0 \to \mathcal{N}_Y \otimes \mathcal{I}_{X|Y} \to \mathcal{N}_Y \to \mathcal{N}_Y \otimes \mathcal{O}_X \to 0$ lead to the following exact sequence

$$(6.2) \qquad 0 \to \mathcal{N}_Y \otimes \mathcal{I}_{X|Y} \to A^1_{X|Y} \to \mathcal{N}_X \to 0.$$

The important ingredient in our proofs is the natural "liaison-invariance" of $A^1_{X|Y}$ "over $\mathcal{N}_Y$" induced by taking the $\mathcal{O}_Y$-dual $(-)^* = \mathcal{H}om_{\mathcal{O}_Y}(-, \mathcal{O}_Y)$, i.e. the isomorphism (cf. [**38**] Corollary 2.12)

$$(6.3) \qquad \begin{array}{ccc} A^1_{X|Y} & \to & \mathcal{N}_Y \\ \wr \| & \circ & \| \\ A^1_{X'|Y} & \to & \mathcal{N}_Y. \end{array}$$

To see this natural isomorphism, let $D(p,q)$ be the Hilbert flag scheme (incidence correspondence) parameterizing "pairs" $X \subset Y$ of closed, equidimensional, locally Cohen-Macaulay subschemes of $\mathbb{P}^N$ with Hilbert polynomial $p$ and $q$ respectively. We see easily that, in the case where $p$ and $q$ are polynomials of the same degree, $q$ is the Hilbert polynomial of a complete intersection of type/degree $f_1, \dots, f_c$ ($Y$ locally Gorenstein is essentially enough to get the isomorphism below) and $p'(v) = q(v) - (-1)^{N-c} p(\Sigma f_i - N - 1 - v)$ is the Hilbert polynomial of the linked object $X' \subset Y$, then there is an isomorphism

$$D(p,q) \cong D(p',q)$$

as sets, defined by sending $X \subset Y$ to $X' \subset Y$ (provided we restrict $D(p,q)$ and $D(p',q)$ to the open part of pairs where $Y$ is a global complete intersection). In [**38**], Theorem 2.6, this bijection is shown to be an isomorphism as schemes (over $\mathrm{Hilb}^q(\mathbb{P}^N)$). The corresponding isomorphism of tangent spaces is just $H^0(A^1_{X|Y}) \cong H^0(A^1_{X'|Y})$ "over $H^0(\mathcal{N}_Y)$," and we get $A^1_{X|Y} \cong A^1_{X'|Y}$ "over $\mathcal{N}_Y$" by "sheafifying" the argument (i.e. by the corresponding argument for local $k$-algebras; it does not matter for the conclusion that we lose the representability of the functor which corresponds to $D(p,q)$); cf. [**38**], Section 2, for details.

The geometric approach indicated above leads easily to the liaison-invariance of most $H^i_*(\mathcal{N}_X)$ and, after a more detailed study, to the liaison-invariance of a certain kernel/cokernel involving $H^i_*(\mathcal{N}_X)$ for the remaining $i > 0$. Only the liaison-invariance of this kernel and cokernel is partially new compared to the treatise [**14**]. Under some local complete intersection (l.c.i.) requirements one may, however, get the invariance of this kernel and cokernel from [**14**], Theorem 2.3, using duality, while our approach has the liaison-invariance of $H^i_{\mathfrak{m}}(\mathrm{Ext}^c_R(R/I_X, R) \otimes_R I_X)$ as a corollary under certain locally licci (e.g. l.c.i.) requirements which [**14**] does not require. Let $A = R/I_X$. In particular we see the Cohen-Macaulayness of $K_A \otimes_A I_X/I_X^2$ for licci subschemes [**13**], and we get the vanishing of the algebra cohomology group $H^2(R,A,A)$ under slightly weaker conditions than those given in [**35**] or [**14**]. We also get an interesting duality involving $H^i_{\mathfrak{m}}(K_A \otimes_R I_X)$, $0 \le i \le \dim X$, for locally Gorenstein subschemes of codimension 3. In the local complete intersection case, $H^i_{\mathfrak{m}}(K_A(N+1) \otimes_R I_X) \cong H^{i+2}(R,A,A)$, and we therefore get a duality for $H^i(R,A,A)$. Moreover, even though our methods in this chapter are fully developed in [**38**], we succeed in generalizing some results of [**38**] which required geometric linkage, to algebraic linkage. Since [**38**] will serve as a standard reference, we take the opportunity to mention the following misprints in [**38**]:

- $H^1(\mathcal{I}_{X|Y}(f_i + v))$ and $H^1(f_i + v))$ in (2.19.1) should be $H^1(\mathcal{I}_X(f_i + v))$ and $H^1(\mathcal{O}_X(f_i + v))$, respectively
- $H^1(R,A,A)$ (resp. $H^2(R,A,A)$) in Lemma 4.7 (resp. Remark 4.9) should be $H^1(R,A,\bar{A})$ (resp. $H^2(R,A,\bar{A})$)

Several liaison results of this shapter consider the following natural map

$$\delta_X : H^n_*(\mathcal{N}_X) \to \mathrm{Hom}_R(I_X, H^{n+1}_{\mathfrak{m}}(A))$$

(where $n = \dim X$), which one may define using the first terms of an $R$-free resolution of $I_X$

$$\cdots \to \bigoplus_i R(-n_{2,i}) \to \bigoplus_i R(-n_{1,i}) \to I_X \to 0.$$

Since $\mathcal{N}_X = \mathcal{H}om(\mathcal{I}_X, \mathcal{O}_X)$, we get an exact sequence

$$0 \to \mathcal{N}_X \to \bigoplus_i \mathcal{O}_X(n_{1,i}) \to \bigoplus_i \mathcal{O}_X(n_{2,i})$$

and the existence of $\delta_X$ follows by taking cohomology because $H_*^n(\mathcal{O}_X) \cong H_{\mathfrak{m}}^{n+1}(A)$ (for $\dim X > 0$). It is easy to see that $\delta_X$ is an isomorphism provided $I_X/I_X^2$ is a free $A = R/I_X$ module and $\dim X > 0$. In the case $\dim X = 0$, $\delta_X$ is well-defined and it is surjective if $I_X/I_X^2$ is a free $A$ module. We will prove

THEOREM 6.1. *Let $X \subset \mathbb{P}^N$ and $X' \subset \mathbb{P}^N$ (with saturated ideals $I_X$ and $I_{X'}$ of $R$ resp.) be ACM schemes, algebraically linked by some complete intersection $Y \subset \mathbb{P}^N$ of dimension $n$. If $n > 0$, then, as graded $R$-modules we have*

$$H_*^i(\mathcal{N}_X) \cong H_*^i(\mathcal{N}_{X'}) \quad \textit{for } 1 \le i \le n-1, \qquad \textit{and}$$

$$\ker \delta_X \cong \ker \delta_{X'}.$$

*Moreover, at least as graded $k$-modules, we have for any $n \ge 0$*

$$\operatorname{coker} \delta_X \cong \operatorname{coker} \delta_{X'}.$$

REMARK 6.2. See Proposition 9.20 for the relation between $H^0(\mathcal{N}_X)$ and $H^0(\mathcal{N}_{X'})$.

REMARK 6.3. (i) Under some extra conditions, $\ker \delta_X$, $\operatorname{coker} \delta_X$ and $H_*^i(\mathcal{N}_X)$ are isomorphic to certain algebra cohomology groups $H^j(R, A, A)$. Indeed let $j$ be an integer, $2 \le j \le n+2$. Using for instance [**40**] (Lemma 3.5 and proof of Theorem 3.6), one may see

$$H_{\mathfrak{m}}^0(H^j(R, A, A)) \cong \ker \delta_X \text{ (resp. } \cong \operatorname{coker} \delta_X) \text{ if } j = n+1 \text{ (resp. } j = n+2)$$

$$H_{\mathfrak{m}}^0(H^j(R, A, A)) \cong H_*^{j-1}(\mathcal{N}_X) \text{ if } 2 \le j \le n$$

provided the sheaf $H^i(R, A, A)^{\sim} = 0$ for $2 \le i < j$. Therefore we get the liaison-invariance of $H^i(R, A, A)$ for $2 \le i \le j$ from Theorem 6.1 if we suppose $H^i(R, A, A)^{\sim} = H^i(R, A', A')^{\sim} = 0$ for $2 \le i \le j$ (for instance if $X$ and $X'$ are local complete intersections in $\mathbb{P}^N$); cf.[**14**], Theorem 4.1.

(ii) We expect a more functorial proof will show the isomorphism of the cokernels of Theorem 6.1 to be $R$-linear as well. If the linkage is geometric, we can prove the $R$-linearity as indicated in the proof for $n > 0$, while the case $n = 0$ follows for instance from [**38**], Theorem 4.2.

We will prove Theorem 6.1 shortly (on page 40). First, we will give an alternative description of $A_{X|Y}^1$ from which we easily see $A_{X|Y}^1 \cong A_{X'|Y}^1$. Indeed, to prove the commutativity of a certain diagram appearing in the proof of the liaison-invariance of $\ker \delta_X$ and $\operatorname{coker} \delta_X$, it is desirable to describe an isomorphism $A_{X|Y}^1 \cong A_{X'|Y}^1$, fitting into the commutative diagram (6.3), directly. To do this, let $I_{X|Y} = I_X/I_Y$ and let

$$(6.4) \qquad \begin{array}{ccccccc} P_1 & \to & P_0 & \to & I_{X|Y} & \to & 0, \quad \text{and} \\ Q_1 & \to & Q_0 & \to & I_{X'|Y} & \to & 0 \end{array}$$

be exact sequences, where $P_i$ and $Q_i$ are $B$-free, and let $Z = \ker(P_0 \to I_{X|Y})$. Applying $(-)^* = \mathrm{Hom}_B(-, B)$, we get

$$\begin{array}{llllllll} 0 & \to & I^*_{X'|Y} = A & \to & Q_0^* & \to & Q_1^*, & \text{and} \\ 0 & \to & I^*_{X|Y} = A' & \to & P_0^* & \to & P_1^*. & \end{array} \tag{6.5}$$

Consider

$$\begin{array}{ccccccc} & & 0 & & 0 & & 0 \\ & & \downarrow & & \downarrow & & \downarrow \\ 0 & \to & \mathrm{Hom}_B(I_{X|Y}, A) & \to & \mathrm{Hom}_B(P_0, A) & \xrightarrow{d_A} & \mathrm{Hom}_B(P_1, A) \\ & & \downarrow & & \downarrow & & \downarrow \delta_1 \\ 0 & \to & \mathrm{Hom}_B(I_{X|Y}, Q_0^*) & \to & \mathrm{Hom}_B(P_0, Q_0^*) & \xrightarrow{d_0} & \mathrm{Hom}_B(P_1, Q_0^*) \\ & & \downarrow & & \downarrow \delta_0 & & \downarrow \\ 0 & \to & \mathrm{Hom}_B(I_{X|Y}, Q_1^*) & \to & \mathrm{Hom}_B(P_0, Q_1^*) & \to & \mathrm{Hom}_B(P_1, Q_1^*). \end{array} \tag{6.6}$$

Then we can find the first terms of a "resolution of $\mathrm{Hom}_B(I_{X|Y}, A)$ by the double complex" as the upper horizontal sequence in the commutative diagram

(6.7)

$$\begin{array}{cccccccc} \ker(\delta_0, d_0) & \hookrightarrow & \mathrm{Hom}_B(P_0, Q_0^*) & \xrightarrow{(\delta_0, d_0)} & \mathrm{Hom}_B(P_0, Q_1^*) \oplus \mathrm{Hom}_B(P_1, Q_0^*) & & & \\ \uparrow & & \circlearrowleft & & \uparrow (0, \delta_1) & & & \\ \ker d_A & \hookrightarrow & \mathrm{Hom}_B(P_0, A) & \xrightarrow{d_A} & \mathrm{Hom}_B(P_1, A) & & & \\ \| & & \| & & \circlearrowleft & & & \\ \mathrm{Hom}_B(I_{X|Y}, A) & \hookrightarrow & \mathrm{Hom}_B(P_0, A) & \to & \mathrm{Hom}_B(Z, A) & \to & \mathrm{Ext}^1_B(I_{X|Y}, A) & \to 0. \end{array}$$

Indeed, we leave it as an easy exercise to see that

$$\mathrm{Hom}_B(I_{X|Y}, A) \xrightarrow{\sim} \ker(\delta_0, d_0)$$

and the left injection of

$$\mathrm{Ext}^1_B(I_{X|Y}, A) \hookrightarrow \operatorname{coker} d_A \hookrightarrow \operatorname{coker}(\delta_0, d_0). \tag{6.8}$$

To see the right injection, take $x_1 \in \mathrm{Hom}_B(P_1, A)$ such that $(0, \delta_1)(x_1) \in \mathrm{im}(\delta_0, d_0)$, i.e. $(0, \delta_1(x_1)) = (\delta_0(y), d_0(y))$ for some $y \in \mathrm{Hom}_B(P_0, Q_0^*)$. Since $\delta_0(y) = 0$, then (6.6) shows that there exists $x_0 \in \mathrm{Hom}_B(P_0, A)$ which maps to $y$ in $\mathrm{Hom}_B(P_0, Q_0^*)$ and, by the injectivity of $\delta_1$, to $d_A(x_0) = x_1$ in $\mathrm{Hom}_B(P_1, A)$. Hence $x_1$ maps to zero in $\operatorname{coker} d_A$, which shows the injectivity.

Now we consider the exact sequence

$$0 \to I_Y \otimes_R A \cong I_Y/I_Y I_X \to I_X/I_Y I_X \to I_{X|Y} \to 0 \tag{6.9}$$

which we may assume fits into a commutative diagram

$$\begin{array}{ccccccccc} 0 & \to & I_Y \otimes_R A & \to & I_X \otimes_R B & \to & I_{X|Y} & \to & 0 \\ & & \uparrow & & \uparrow & & \| & & \\ 0 & \to & Z & \to & P_0 & \to & I_{X|Y} & \to & 0 \end{array} \tag{6.10}$$

where the center vertical map is an epimorphism. Applying $\widetilde{\mathrm{Hom}_B(-,A)}$ to (6.10), we get two of the horizontal exact sequences in the commutative diagram

(6.11)

$$\begin{array}{ccccccccc}
 & & A^1_{X|Y} & \to & \mathcal{N}_Y & & & & \\
 & & \downarrow & & \downarrow & & & & \\
\widetilde{\mathrm{Hom}_B(I_{X|Y},A)} & \hookrightarrow & \mathcal{N}_X & \to & \mathcal{N}_Y \otimes \mathcal{O}_X & \to & \widetilde{\mathrm{Ext}^1_B(I_{X|Y},A)} & & \\
\| & & \downarrow & & \downarrow & & \| & & \\
\widetilde{\mathrm{Hom}_B(I_{X|Y},A)} & \hookrightarrow & \widetilde{\mathrm{Hom}_B(P_0,A)} & \to & \widetilde{\mathrm{Hom}_B(Z,A)} & \to & \widetilde{\mathrm{Ext}^1_B(I_{X|Y},A)} & \to & 0 \\
\downarrow \wr & & \downarrow & & \downarrow & & \hookdownarrow & & \\
\widetilde{\ker(\delta_0,d_0)} & \hookrightarrow & \widetilde{\mathrm{Hom}_B(P_0,Q_0^*)} & \to & \begin{pmatrix} \widetilde{\mathrm{Hom}_B(P_0,Q_1^*)} \\ \oplus \\ \widetilde{\mathrm{Hom}_B(P_1,Q_0^*)} \end{pmatrix} & \to & \widetilde{\mathrm{coker}(\delta_0,d_0)} & \to & 0.
\end{array}$$

The injectivity we proved in (6.8) readily implies that the diagram, whose morphisms are obtained by taking obvious compositions of maps in (6.11)

$$(6.12)\qquad \begin{array}{ccc}
A^1_{X|Y} & \to & \mathcal{N}_Y \\
\downarrow & \square & \downarrow \\
\widetilde{\mathrm{Hom}_B(P_0,Q_0^*)} & \xrightarrow{(\delta_0,d_0)} & \widetilde{\mathrm{Hom}_B(P_0,Q_1^*)} \oplus \widetilde{\mathrm{Hom}_B(P_1,Q_0^*)}
\end{array}$$

is *cartesian*. We have a corresponding diagram for $A^1_{X'|Y}$ in which the lower horizontal map is the sheafification of $(d_0', \delta_0')$ in the exact sequence

(6.13)

$$0 \to \mathrm{Hom}_B(I_{X'|Y}, A') \to \mathrm{Hom}_B(Q_0, P_0^*) \xrightarrow{(d_0',\delta_0')} \mathrm{Hom}_B(Q_0, P_1^*) \oplus \mathrm{Hom}_B(Q_1, P_0^*)$$

given by (6.7). Using that

$$(6.14)\qquad \mathrm{Hom}_B(P,Q) \cong \mathrm{Hom}_B(Q^*, P^*)$$

for $B$-free modules $P$ and $Q$, and in particular that the composition

$$\mathcal{N}_Y = \widetilde{\mathrm{Hom}_B(I_Y/I_Y^2, B)} \cong \widetilde{\mathrm{Hom}_B(B^*, (I_Y/I_Y^2)^*)} \cong \widetilde{(I_Y/I_Y^2)^*}$$

really is the identity, we deduce from (6.12) the existence of an isomorphism $A^1_{X|Y} \cong A^1_{X'|Y}$ fitting into (6.3). Of course (6.7), (6.13) and (6.14) induce isomorphisms

$$\begin{array}{rcll}
\mathrm{Hom}_B(I_{X|Y}, A) & \cong & \mathrm{Hom}_B(I_{X'|Y}, A'), & \text{and} \\
\mathrm{coker}(\delta_0, d_0) & \cong & \mathrm{coker}(d_0', \delta_0') &
\end{array}$$

where the latter one commutes (cf. (6.8)) with the isomorphism

$$\mathrm{Ext}^1_B(I_{X|Y}, A) \xrightarrow{\sim} \mathrm{Ext}^1_B(I_{X'|Y}, A')$$

which we get by applying $(-)^*$ to an extension

$$e: \quad 0 \to A \to E \to I_{X|Y} \to 0.$$

Finally, we remark that in a similar manner to (6.7) we can find the first terms in a "resolution of $\mathrm{Hom}_B(I_{X|Y}, H^{n+1}_{\mathfrak{m}}(A))$." Indeed, we claim that we have exact vertical sequences in

(6.15)
$$\begin{array}{ccc}
\mathrm{Hom}_B(I_{X|Y}, H^{n+1}_{\mathfrak{m}}(A)) & = & \mathrm{Hom}_B(I_{X|Y}, H^{n+1}_{\mathfrak{m}}(A)) \\
\cap\!\downarrow & & \cap\!\downarrow \\
\mathrm{Hom}_B(P_0, H^{n+1}_{\mathfrak{m}}(A)) & \hookrightarrow & \mathrm{Hom}_B(P_0, H^{n+1}_{\mathfrak{m}}(Q_0^*)) \\
\downarrow & & \downarrow (\delta_0^h, d_0^h) \\
\mathrm{Hom}_B(Z, H^{n+1}_{\mathfrak{m}}(A)) & \to & \mathrm{Hom}_B(P_0, H^{n+1}_{\mathfrak{m}}(Q_1^*)) \oplus \mathrm{Hom}_B(P_1, H^{n+1}_{\mathfrak{m}}(Q_0^*))
\end{array}$$

as well as an injection

$$\mathrm{Ext}^1_B(I_{X|Y}, H^{n+1}_{\mathfrak{m}}(A)) \hookrightarrow \mathrm{coker}(\delta_0^h, d_0^h). \tag{6.16}$$

The latter injection follows by the arguments of (6.8). The proof of (6.15) is the same as (6.7) once we have shown that $H^{n+1}_{\mathfrak{m}}(-)$ is left-exact on (6.5). This left-exactness follows from Gorenstein duality and by applying $\mathrm{Hom}_B(-, H^{n+1}_{\mathfrak{m}}(B))$ to (6.4). Note that the rightmost vertical sequence of (6.15) and its analogue for the linked object, and

$$\mathrm{Hom}_B(P, H^{n+1}_{\mathfrak{m}}(Q)) \cong \mathrm{Hom}_B(Q^*, H^{n+1}_{\mathfrak{m}}(P^*))$$

imply

$$\mathrm{Hom}_B(I_{X|Y}, H^{n+1}_{\mathfrak{m}}(A)) \cong \mathrm{Hom}_B(I_{X'|Y}, H^{n+1}_{\mathfrak{m}}(A')) \tag{6.17}$$

and an isomorphism from $\mathrm{coker}(\delta_0^h, d_0^h)$ onto the corresponding cokernel of the linked object.

Proof. (of Theorem 6.1): Since $Y$ is a complete intersection of type $f_1, \ldots, f_c$, we have $\mathcal{N}_Y \cong \bigoplus_i \mathcal{O}_Y(f_i)$ and the following exact sequence

$$0 \to \bigoplus_i \mathcal{I}_{X|Y}(f_i) \to A^1_{X|Y} \to \mathcal{N}_X \to 0 \tag{6.18}$$

from (6.2). Using

$$0 \to \mathcal{I}_Y \to \mathcal{I}_X \to \mathcal{I}_{X|Y} \to 0$$

and the ACM property of $X$ and $Y$, we get $H^i(\mathcal{I}_{X|Y}(v)) = 0$ for $1 \le i < n$. Hence, for any $v$, we have

$$H^i(A^1_{X|Y}(v)) \cong H^i(\mathcal{N}_X(v)) \text{ for } 1 \le i < n-1 \tag{6.19}$$

and we get the isomorphisms of the theorem in the range $1 \le i < n-1$. Note that the sequence (6.18) and the corresponding sequence twisted by $g$ fit naturally into a commutative diagram by the multiplication of any $G$ in $R$ of degree $g$, from which we see that the isomorphisms of the theorem are $R$-linear. Moreover if $n \ge 2$ there is a long exact sequence and a commutative diagram

$$
\begin{array}{ccc}
0 & & \\
\downarrow & & \\
H^{n-1}(A^1_{X|Y}(v)) & & \\
\downarrow & & \\
H^{n-1}(\mathcal{N}_X(v)) & & \\
\downarrow & & \\
\bigoplus_i H^n(\mathcal{I}_{X|Y}(f_i+v)) & \hookrightarrow & \bigoplus_i H^n(\mathcal{O}_Y(f_i+v)) \\
\downarrow & \circ & \downarrow \wr \\
H^n(A^1_{X|Y}(v)) & \xrightarrow{\phi_{X|Y}} & H^n(\mathcal{N}_Y(v)) \\
\downarrow & & \\
H^n(\mathcal{N}_X(v)) & & \\
\downarrow & & \\
0 & &
\end{array}
\tag{6.20}
$$

where the horizontal injective map

$$\bigoplus_i H^n(\mathcal{I}_{X|Y}(f_i+v)) \hookrightarrow \bigoplus_i H^{n+1}(\mathcal{I}_Y(f_i+v)) \cong \bigoplus_i H^n(\mathcal{O}_Y(f_i+v))$$

and the lower horizontal map $\phi_{X|Y}$, given by (6.3), commute, and we easily get the isomorphism (6.19) in the case $i = n-1$ as well.

If $n \geq 1$, we have the diagram above except for the 0 at the very top. Note that the cokernel of the top horizontal map is precisely $H^n(\mathcal{N}_Y \otimes \mathcal{O}_X(v))$ and therefore, by the Five Lemma, we get the lefthand vertical exact sequence in the following diagram

$$
\begin{array}{ccc}
0 & & 0 \\
\downarrow & & \downarrow \\
\ker \phi_{X|Y} & \xrightarrow{\alpha_{X|Y}} & {}_v\operatorname{Hom}_B(I_{X|Y}, H^{n+1}_{\mathfrak{m}}(A)) \\
\downarrow & & \downarrow \\
H^n(\mathcal{N}_X(v)) & \xrightarrow{\delta_X} & {}_v\operatorname{Hom}_R(I_X, H^{n+1}_{\mathfrak{m}}(A)) \\
\downarrow & \circ & \downarrow \\
H^n(\mathcal{N}_Y \otimes \mathcal{O}_X(v)) & \xrightarrow{\sim} & {}_v\operatorname{Hom}_R(I_Y, H^{n+1}_{\mathfrak{m}}(A)) \\
\downarrow & & \downarrow \\
\operatorname{coker} \phi_{X|Y} & \xrightarrow{\beta_{X|Y}} & {}_v\operatorname{Ext}^1_B(I_{X|Y}, H^{n+1}_{\mathfrak{m}}(A)) \\
\downarrow & & \\
0 & &
\end{array}
\tag{6.21}
$$

where $B = R/I_Y$. The righthand vertical sequence is seen to be exact by applying $\operatorname{Hom}_R(-, H^{n+1}_{\mathfrak{m}}(A))$ to (6.9). Once having understood the naturalness of the horizontal maps, we see that the third horizontal map is indeed an isomorphism because $n > 0$. Since it obviously commutes with $\delta_X$ by the definition of $\delta_X$, the maps given by the top and bottom arrow, $\alpha_{X|Y}$ and $\beta_{X|Y}$, exist. Let

$$\beta'_{X|Y} : \operatorname{coker} \phi_{X|Y} \to \operatorname{coker}(\delta^h_0, d^h_0)$$

be the composition of $\beta_{X|Y}$ and the map

$${}_v\operatorname{Ext}^1_B(I_{X|Y}, H^{n+1}_{\mathfrak{m}}(A)) \hookrightarrow \operatorname{coker}(\delta^h_0, d^h_0)$$

given by (6.16).

*Claim:* The maps $\alpha_{X|Y}$ and $\beta'_{X|Y}$ are "liaison-invariant."

Indeed, by (6.11) and (6.15) and the definition of $\delta_X$, we have a commutative diagram

$$\begin{array}{ccccccc}
0 & & & & & & 0\\
\downarrow & & & & & & \downarrow\\
\ker\phi_{X|Y} & & & \xrightarrow{\quad\alpha_{X|Y}\quad} & & & {}_v\operatorname{Hom}_B(I_{X|Y}, H^{n+1}_{\mathfrak m}(A))\\
\downarrow & & & & & & \downarrow\\
H^n(A^1_{X|Y}(v)) & \longrightarrow & H^n(\mathcal{N}_X(v)) & \longrightarrow & {}_vH^{n+1}_{\mathfrak m}(\operatorname{Hom}_B(P_0,Q_0^*)) & \xrightarrow{\sim} & {}_v\operatorname{Hom}_B(P_0, H^{n+1}_{\mathfrak m}(Q_0^*))\\
\downarrow & & \downarrow & & \downarrow & & \downarrow\\
H^n(\mathcal{N}_Y(v)) & \to & H^n(\mathcal{N}_Y\otimes\mathcal{O}_X(v)) & \to & {}_vH^{n+1}_{\mathfrak m}\begin{pmatrix}\operatorname{Hom}_B(P_0,Q_1^*)\\ \oplus\\ \operatorname{Hom}_B(P_1,Q_0^*)\end{pmatrix} & \xrightarrow{\sim} & \begin{pmatrix}{}_v\operatorname{Hom}_B(P_0,H^{n+1}_{\mathfrak m}(Q_1^*))\\ \oplus\\ {}_v\operatorname{Hom}_B(P_1,H^{n+1}_{\mathfrak m}(Q_0^*))\end{pmatrix}\\
\downarrow & & & & & & \downarrow\\
\operatorname{coker}\phi_{X|Y} & & & \xrightarrow{\quad\beta'_{X|Y}\quad} & & & \operatorname{coker}(\delta_0^h, d_0^h).
\end{array}$$

We have isomorphisms $\ker\phi_{X|Y} \cong \ker\phi_{X'|Y}$ induced by (6.3) (using the explicit description of $A^1_{X|Y} \cong A^1_{X'|Y}$ given by (6.12), (6.13) and (6.14)), and hence it commutes with $H^n(A^1_{X|Y}(v)) \cong H^n(A^1_{X'|Y}(v))$. Similarly, (6.17) commutes with $\operatorname{Hom}_B(P_0, H^{n+1}_{\mathfrak m}(Q_0^*)) \cong \operatorname{Hom}_B(Q_0, H^{n+1}_{\mathfrak m}(P_0^*))$. Therefore $\alpha_{X|Y}$ is "liaison-invariant" because the diagram

$$\begin{array}{ccc}
H^n(A^1_{X|Y}(v)) & \cong & H^n(A^1_{X'|Y}(v))\\
\downarrow & & \downarrow\\
{}_v\operatorname{Hom}_B(P_0, H^{n+1}_{\mathfrak m}(Q_0^*)) & \cong & {}_v\operatorname{Hom}_B(Q_0, H^{n+1}_{\mathfrak m}(P_0^*))
\end{array}$$

commutes by our explicit description of $A^1_{X|Y} \cong A^1_{X'|Y}$. The liaison invariance of $\beta'_{X|Y}$ is shown in a similar way, and the claim is proved.

It follows from the claim that $\ker\alpha_{X|Y}$ is "liaison-invariant". Since $\ker\alpha_{X|Y} \cong \ker\delta_X$ by (6.21), the second isomorphism of the Theorem is proved.

If the linkage is geometric, we can use the fact that $\dim X \cap X' < \dim X$ to prove $\operatorname{coker}\phi_{X|Y} = 0$ from which we get that $\operatorname{coker}\delta_X \cong \operatorname{coker}\alpha_{X|Y}$ is "liaison-invariant." However, for algebraic linkage, we can at least prove $\dim_k \operatorname{coker}\delta_X = \dim_k \operatorname{coker}\delta_{X'}$. Indeed, using (6.21), we get (for $n > 0$) the following exact sequence:

$$0 \to \operatorname{coker}\alpha_{X|Y} \to \operatorname{coker}\delta_X \to \operatorname{coker}\phi_{X|Y} \xrightarrow{\beta_{X|Y}} {}_v\operatorname{Ext}^1_B(I_{X|Y}, H^{n+1}_{\mathfrak m}(A)).$$

Since $\operatorname{coker}\alpha_{X|Y}$ and $\ker\beta_{X|Y} = \ker\beta'_{X|Y}$ are "liaison-invariant" by the proven claim, we conclude easily by the exact sequence above because, as $k$-vector spaces, the direct sum of $\ker\beta_{X|Y}$ and $\operatorname{coker}\alpha_{X|Y}$ is $\operatorname{coker}\delta_X$.

It remains to prove the liaison-invariance of $\operatorname{coker}\delta_X$ provided $n = 0$, a proof we only indicate because, in this case, [**14**], Theorem 2.3, implies the result. (Indeed as argued later in (6.33), one may prove $\operatorname{coker}\delta_X \cong {}_vH^0_{\mathfrak m}(\operatorname{Ext}^2_R(I_X, I_X))$, and conclude by Remark 6.16). If $n = 0$ the isomorphism of (6.21) is only surjective, and the kernel, ${}_v\operatorname{Hom}_R(I_Y, A)$, maps naturally into ${}_v\operatorname{Ext}^1_B(I_{X|Y}, A)$. Now, essentially in the same way as we managed to show that $\operatorname{coker}\phi_{X|Y}$ (resp. the composition

$$\operatorname{coker}\phi_{X|Y} \longrightarrow {}_v\operatorname{Ext}^1_B(I_{X|Y}, H^1_{\mathfrak m}(A)) \hookrightarrow \operatorname{coker}(\delta_0^h, d_0^h))$$

was "liaison-invariant" using (6.1) and (6.3) (resp. mainly (6.12)), the corresponding arguments for the graded deformation functor (cf. [**41**]) imply the "liaison-invariance" of $\operatorname{coker}\phi_{B,A}$ (resp. $\operatorname{coker}\phi_{B,A} \to {}_v\operatorname{Ext}^1_B(I_{X|Y}, A)$ and $\operatorname{coker}\phi_{X|Y} \to H^0(\mathcal{E}xt^1_{\mathcal{O}_Y}(\mathcal{I}_{X|Y}, \mathcal{O}_X(v))))$, where $\phi_{B,A} : A^1_{B,A} \to {}_v\operatorname{Hom}_R(I_Y, B)$ is natural and $A^1_{B,A}$ is given by the cartesian diagram (6.1) where we have replaced sheaf $\mathcal{H}om(-,-)$ by the corresponding graded piece ${}_v\operatorname{Hom}(-,-)$ of the graded $\operatorname{Hom}(-,-)$.

If $\gamma : \operatorname{im}\delta_X \to {}_v\operatorname{Hom}_R(I_Y, H^1_{\mathfrak{m}}(A))$, then one proves the exactness of the sequences

$$0 \to {}_v\operatorname{Hom}_B(I_{X|Y}, A) \to H^0(\mathcal{N}_{X|Y}(v)) \to \ker\gamma \to \operatorname{coker}\phi_{B,A} \\ \to \operatorname{coker}\phi_{X|Y} \to \operatorname{coker}\gamma \to 0$$

$$0 \to \ker\gamma \to {}_v\operatorname{Hom}_B(I_{X|Y}, H^1_{\mathfrak{m}}(A)) \to \operatorname{coker}\delta_X \to \operatorname{coker}\gamma \\ \to {}_v\operatorname{Ext}^1_B(I_{X|Y}, H^1_{\mathfrak{m}}(A)).$$

Since the first, second, fourth and fifth groups in the first sequence, and the morphism above between the two latter ones, are "liaison-invariant," we get that $\ker\gamma$ and $\operatorname{coker}\gamma$ are "liaison-invariant," at least as $k$-vector spaces. Applying exactly the same argument to the second exact sequence, we get the "liaison-invariance" of $\operatorname{coker}\delta_X$, and we are done. □

REMARK 6.4. An interesting observation from the proof above is that we get the "liaison-invariance" of the groups $H^i(A^1_{X|Y})$ also for schemes $X$ which are not ACM and even for $Y$'s which are not a complete intersection (e.g. (6.3) holds if $Y$ is arithmetically Gorenstein, while, however, the formulation of (6.1) and (6.2) above require $Y$ to be a local complete intersection). The groups $H^i(A^1_{X|Y})$ which usually depend on $Y$, fit into the long exact sequence induced by (6.2), giving interesting information of the cohomology $H^i(\mathcal{N}_X(v))$ in terms of $H^i(\mathcal{N}_{X'}(v))$ in general, cf. [**38**], (2.19.1). In particular (6.2) implies the "liaison-invariance" of $\chi(\mathcal{N}_X(v)) + \sum\chi(\mathcal{I}_{X|Y}(f_i + v))$, hence of $\chi(\mathcal{N}_X(v)) - \sum\chi(\mathcal{O}_X(f_i + v))$, for equidimensional, locally Cohen-Macaulay subschemes of $\mathbb{P}^N$, linked via a complete intersection of type $f_1, \dots, f_c$. As an application recall that, for smooth curves $C$ in $\mathbb{P}^N$ of degree $d$ and genus $g$, one knows

$$\chi(\mathcal{N}_C(v)) = (N+1)d + (N-3)(1-g) + (N-1)dv. \tag{6.22}$$

The "liaison-invariant" expression $\chi(\mathcal{N}_C(v)) - \sum\chi(\mathcal{O}_C(f_i+v))$ is therefore equal to $2g - 2 - d(\sum f_i - N - 1)$, which one knows is equal to the corresponding formula for the linked curve by Proposition 2.5. Hence (6.22) holds for any curve of degree $d$ and arithmetic genus $g$ in the linkage class of a smooth curve. This generalizes [**38**], Corollary 2.17.

When we defined $\delta_X$ we observed that $\delta_X$ was an isomorphism for complete intersections. Since it is well known that complete intersections satisfy $H^i_*(\mathcal{N}_X) = 0$ for $1 \le i < \dim X$, Theorem 6.1 immediately implies the following result which partially generalizes [**39**], Corollary 1.3, and [**44**], Theorem 2.6.

COROLLARY 6.5. *If $X \subset \mathbb{P}^N$ (of dimension $n$) is licci, then*

$$H^i(\mathcal{N}_X(v)) = 0 \qquad \text{for } 1 \le i \le n-1 \text{ and any } v.$$

*Moreover the natural map*

$$\delta_X : H^n_*(\mathcal{N}_X) \to \operatorname{Hom}_R(I_X, H^{n+1}_{\mathfrak{m}}(R/I_X))$$

*is an isomorphism (resp. surjection) provided* $n \geq 1$ *(resp.* $n = 0$*).*

To get easily some later results, we recall that the right derived functors ${}_v\operatorname{Ext}^i_{\mathfrak{m}}(M, -)$ of ${}_vH^0_{\mathfrak{m}}(\operatorname{Hom}_R(M, -))$ are equipped with spectral sequences

$$E_2^{p,q} = {}_v\operatorname{Ext}^p_R(M_1, H^q_{\mathfrak{m}}(M_2)) \Rightarrow {}_v\operatorname{Ext}^{p+q}_{\mathfrak{m}}(M_1, M_2) \text{ and} \tag{6.23}$$

$$E_2^{p,q} = {}_vH^p_{\mathfrak{m}}(\operatorname{Ext}^q_R(M_1, M_2)) \Rightarrow {}_v\operatorname{Ext}^{p+q}_{\mathfrak{m}}(M_1, M_2) \tag{6.24}$$

($\Rightarrow$ means "converging to" and $M_i$ are graded $R$-modules of finite type) and a long exact sequence ([**28**], exp. VI, or [**29**])

(6.25)

$$\to {}_v\operatorname{Ext}^i_{\mathfrak{m}}(M_1, M_2) \to {}_v\operatorname{Ext}^i_R(M_1, M_2) \to \operatorname{Ext}^i_{\mathcal{O}_{\mathbb{P}}}(\tilde{M_1}, \tilde{M_2}(v)) \to {}_v\operatorname{Ext}^{i+1}_{\mathfrak{m}}(M_1, M_2) \to$$

(where $\mathbb{P} = \mathbb{P}^N$) which sometimes is efficient to combine with the duality isomorphism;

$${}_v\operatorname{Ext}^i_{\mathfrak{m}}(N_2, N_1) \cong {}_{-v-N-1}\operatorname{Ext}^{N+1-i}_R(N_1, N_2)^\vee \tag{6.26}$$

(valid for any integers $i$ and $v$, cf. [**39**], Theorem 1.1), even though in this paper (where we usually consider ACM subschemes) we can reach our conclusions by using the functoriality of Serre-Gorenstein duality to appropriate exact sequences, without too much extra work.

A coherent non-trivial $\mathcal{O}_X$-module $\mathcal{F}$ is said to be a *maximal Cohen-Macaulay* $\mathcal{O}_X$*-module* if $\mathcal{E}xt^i_{\mathcal{O}_X}(\mathcal{F}, \omega_X) = 0$ for $i > 0$. The corresponding vanishing for a finitely generated module $M \neq 0$ over a local Cohen-Macaulay ring $(A, \mathfrak{m})$ is by Gorenstein duality equivalent to the vanishing of $H^i_{\mathfrak{m}}(M)$ for $i < dim A$, i.e. to $M$ being a maximal Cohen-Macaulay $A$-module. Note that if $\mathcal{F}$ is locally free then it is a maximal Cohen-Macaulay $\mathcal{O}_X$-module, and that $\mathcal{I}_X/\mathcal{I}_X^2 \otimes_{\mathcal{O}_X} \omega_X$ will be a maximal Cohen-Macaulay $\mathcal{O}_X$-module if $X$ is a local complete intersection.

COROLLARY 6.6. *Let* $X \subset \mathbb{P}^N$, $n = \dim X$, *be an ACM subscheme of codimension* $c$ *belonging to the same CI-liaison class as* $X' \subset \mathbb{P}^N$, *and let* $K_A$ *and* $K_{A'}$ *be the canonical modules of* $A = R/I_X$ *and* $A' = R/I_{X'}$ *respectively. Assume that*

- $\mathcal{I}_X/\mathcal{I}_X^2 \otimes_{\mathcal{O}_X} \omega_X$ *is a maximal Cohen-Macaulay* $\mathcal{O}_X$*-module, and*
- $\mathcal{I}_{X'}/\mathcal{I}_{X'}^2 \otimes_{\mathcal{O}_{X'}} \omega_{X'}$ *is a maximal Cohen-Macaulay* $\mathcal{O}_{X'}$*-module.*

*Then*

$$H^j_{\mathfrak{m}}(K_A \otimes_R I_X) \cong H^j_{\mathfrak{m}}(K_{A'} \otimes_R I_{X'})$$

*are isomorphic as graded* $R$*-modules (resp. as graded* $k$*-modules) provided* $0 < j \leq n$ *(resp.* $0 \leq j \leq n$*). In particular, if* $X' \subset \mathbb{P}^N$ *is a complete intersection, then*

$$H^j_{\mathfrak{m}}(K_A \otimes_A I_X/I_X^2) = 0, \qquad \text{for } 0 \leq j \leq n.$$

PROOF. In general we have (cf. [**32**])

$$\mathcal{N}_X \cong \mathcal{H}om_{\mathcal{O}_X}(\mathcal{I}_X/\mathcal{I}_X^2 \otimes_{\mathcal{O}_X} \omega_X, \omega_X)$$

and

$$H^i_*(\mathcal{N}_X) \cong \operatorname{Ext}^i_{\mathcal{O}_X}(\mathcal{I}_X/\mathcal{I}_X^2 \otimes_{\mathcal{O}_X} \omega_X, \omega_X), \quad 0 \leq i \leq n \tag{6.27}$$

by the Cohen-Macaulayness of $\mathcal{I}_X/\mathcal{I}_X^2 \otimes_{\mathcal{O}_X} \omega_X$ and the spectral sequence relating local $\mathcal{E}xt$ and global Ext. Since $\operatorname{pd}_R A = \operatorname{pd} \mathcal{O}_X = c$, we get

$$\mathcal{E}xt^c_{\mathcal{O}_\mathbb{P}}(\mathcal{O}_X, \mathcal{I}_X) \cong \mathcal{E}xt^c_{\mathcal{O}_\mathbb{P}}(\mathcal{O}_X, \mathcal{O}_\mathbb{P}) \otimes_{\mathcal{O}_\mathbb{P}} \mathcal{I}_X \cong \omega_X(N+1) \otimes_{\mathcal{O}_X} \mathcal{I}_X/\mathcal{I}_X^2$$

and $\operatorname{Ext}^c_R(A, I_X) \cong K_A(N+1) \otimes_R I_X$. If $0 < i \leq n$, we have by Serre duality;

(6.28)

$$H^{i+1}_\mathfrak{m}(\operatorname{Ext}^c_R(A, I_X)) \cong H^i_*(\mathcal{E}xt^c_{\mathcal{O}_\mathbb{P}}(\mathcal{O}_X, \mathcal{I}_X)) \cong H^i_*(\mathcal{I}_X/\mathcal{I}_X^2 \otimes_{\mathcal{O}_X} \omega_X)(N+1)$$

$$\cong \operatorname{Ext}^{n-i}_{\mathcal{O}_X}(\mathcal{I}_X/\mathcal{I}_X^2 \otimes_{\mathcal{O}_X} \omega_X, \omega_X)^\vee(N+1) \cong H^{n-i}_*(\mathcal{N}_X)^\vee(N+1)$$

and $H^{i+1}_\mathfrak{m}(\operatorname{Ext}^c_R(A', I_{X'})) \cong H^{n-i}_*(\mathcal{N}_{X'})^\vee(N+1)$. We conclude by Theorem 6.1 in the case $2 \leq j \leq n$, letting $j = i+1$. For $i = -1$ and 0, we have the exact sequence

(6.29)

$$\begin{aligned} 0 \to H^0_\mathfrak{m}(\operatorname{Ext}^c_R(A, I_X)) \to \operatorname{Ext}^c_R(A, I_X) \quad &\to \quad H^0_*(\omega_X \otimes_{\mathcal{O}_X} \mathcal{I}_X/\mathcal{I}_X^2)(N+1) \\ &\to \quad H^1_\mathfrak{m}(\operatorname{Ext}^c_R(A, I_X)) \to 0. \end{aligned}$$

The twisted dual of the natural morphism in the middle is precisely the natural map

$$\delta_X : H^n_*(\mathcal{N}_X) \to \operatorname{Hom}_R(I_X, H^{n+1}_\mathfrak{m}(A))$$

of Theorem 6.1 by Serre duality and (6.26). (Instead of (6.26) we can apply Gorenstein duality to the first terms of a minimal resolution of $I_X$ and the right exactness of $\operatorname{Ext}^c_R(A, -)$ to see that $\operatorname{Ext}^c_R(A, I_X)^\vee \cong \operatorname{Hom}_R(I_X, H^{n+1}_\mathfrak{m}(A))(-N-1)$). The proof is therefore complete. □

REMARK 6.7. (i) Corollary 6.6 is proved in [**14**], Theorem 2.3, without any Cohen-Macaulay claims on $\mathcal{I}_X/\mathcal{I}_X^2 \otimes_{\mathcal{O}_X} \omega_X$ and $\mathcal{I}_{X'}/\mathcal{I}_{X'}^2 \otimes_{\mathcal{O}_{X'}} \omega_{X'}$; cf. Remark 6.16. Note however that, in our approach, we also easily get the maximal Cohen-Macaulayness of $K_A \otimes I_X/I_X^2$, i.e. the vanishing of $\operatorname{Ext}^i_A(K_A \otimes I_X/I_X^2, K_A)$ for $i > 0$ without the hypothesis $\mathcal{E}xt^i_{\mathcal{O}_X}(\mathcal{I}_X/\mathcal{I}_X^2 \otimes_{\mathcal{O}_X} \omega_X, \omega_X) = 0$ for $i > 0$ by an induction argument on $n = \dim X$. Indeed, in the case $n = 0$, any non-trivial coherent $\mathcal{O}_X$-module is Cohen-Macaulay, and we conclude by induction because (the vanishing in) Corollary 6.6 holds for a local Cohen-Macaulay ring, as well as for a graded ring.

(ii) From (i) we get at once that $\mathcal{E}xt^i_{\mathcal{O}_X}(\mathcal{I}_X/\mathcal{I}_X^2 \otimes_{\mathcal{O}_X} \omega_X, \omega_X) = 0$ for $i > 0$ provided $X$ is locally licci.

(iii) From the proof we also get $H^{n+1}_\mathfrak{m}(K_A \otimes_R I_X) \cong H^0_*(\mathcal{N}_X)^\vee$ for $n \geq 1$.

(iv) We note that these invariants can be readily computed with computer algebra systems such as Macaulay or CoCoA.

For the rest of this chapter we mainly consider subschemes $X$ in $\mathbb{P}^N$ of codimension 3. Then we shall see that $H^i_\mathfrak{m}(\operatorname{Ext}^3_R(R/I_X, I_X))$ is isomorphic to $H^{i+1}_*(\mathcal{N}_X)$ for most $i$, and we get the liaison-invariance of $H^i_\mathfrak{m}(K_{R/I_X} \otimes_R I_X)$ in general and an interesting duality by combining with (6.28) and (6.29). More precisely we have

PROPOSITION 6.8. *Let $X \subset \mathbb{P}^{n+3}$ be an ACM subscheme of codimension 3, and let $K_A$ be the canonical module of $A = R/I_X$. Then*

$$H^i_\mathfrak{m}(K_A \otimes_R I_X)(n+4) \cong H^{i+1}_*(X, \mathcal{N}_X), \qquad \textit{for } 0 \leq i \leq n-2,$$

*as graded R-modules. Moreover for $n \geq 1$ there is an exact sequence*

$$0 \to H_{\mathfrak{m}}^{n-1}(K_A \otimes_R I_X)(n+4) \to H_*^n(\mathcal{N}_X) \xrightarrow{\delta_X} \operatorname{Hom}_R(I_X, H_{\mathfrak{m}}^{n+1}(A)) \to H_{\mathfrak{m}}^n(K_A \otimes_R I_X)(n+4) \to 0,$$

*and for $n = 0$ there is an exact sequence*

$$0 \to \operatorname{Ext}_R^1(I_X, I_X) \to H_*^0(\mathcal{N}_X) \xrightarrow{\delta_X} \operatorname{Hom}_R(I_X, H_{\mathfrak{m}}^1(A)) \to H_{\mathfrak{m}}^0(K_A \otimes_R I_X)(4) \to 0.$$

*In particular $H_{\mathfrak{m}}^i(K_A \otimes_R I_X)$ is invariant under linkage as a graded $R$ (resp. $k$)-module for any $0 \leq i < n$ (resp. $0 \leq i \leq n$), and if $X \subset \mathbb{P}^{n+3}$ is locally Gorenstein, then we have $R$-module isomorphisms*

$$H_{\mathfrak{m}}^i(K_A \otimes_R I_X)(n+4) \cong H_{\mathfrak{m}}^{n-i}(K_A \otimes_R I_X)^\vee, \qquad \textit{for } 0 \leq i \leq n.$$

PROOF. Note that

$$K_A(n+4) \otimes_R I_X \cong \operatorname{Ext}_R^3(A, I_X) \cong \operatorname{Ext}_R^2(I_X, I_X) \tag{6.30}$$

and we build our proof on the last group. Indeed using the spectral sequence (6.23), observing that $H_{\mathfrak{m}}^q(I_X) = 0$ for $q \leq n+1$, we get

$$\operatorname{Ext}_{\mathfrak{m}}^j(I_X, I_X) = 0 \text{ for } j \leq n+1. \tag{6.31}$$

Since the codimension is 3, i.e. $\operatorname{pd}_R(I_X) = 2$, we also have $\operatorname{Ext}_R^j(I_X, I_X) = 0$ for $j \geq 3$ and $\operatorname{Hom}_R(I_X, I_X) \cong R$. Therefore, if $p + q < n+4$ and $M_1 = M_2 = I_X$, the spectral sequence (6.24) consists of at most two non-vanishing terms. Since it converges to zero by (6.31), we have

$$H_{\mathfrak{m}}^i(\operatorname{Ext}_R^2(I_X, I_X)) \cong H_{\mathfrak{m}}^{i+2}(\operatorname{Ext}_R^1(I_X, I_X)) \quad \text{for } i \leq n-2. \tag{6.32}$$

Since $\mathcal{N}_X \cong \mathcal{E}xt^1(\mathcal{I}_X, \mathcal{I}_X)$, we get the first isomorphism of the Proposition in the required ranged $0 \leq i \leq n-2$ by using (6.30).

By the spectral sequence (6.23), $\operatorname{Ext}_{\mathfrak{m}}^{n+2}(I_X, I_X) \cong \operatorname{Hom}_R(I_X, H_{\mathfrak{m}}^{n+1}(A))$. Hence, using the spectral sequence (6.24) we obtain the exactness of

(6.33)
$$0 \to H_{\mathfrak{m}}^{n-1}(\operatorname{Ext}_R^2(I_X, I_X)) \to H_{\mathfrak{m}}^{n+1}(\operatorname{Ext}_R^1(I_X, I_X)) \to \operatorname{Hom}_R(I_X, H_{\mathfrak{m}}^{n+1}(A)) \to H_{\mathfrak{m}}^n(\operatorname{Ext}_R^2(I_X, I_X)) \to 0$$

provided $n \geq 1$. The zero to the right is by reasons of dimension, and we get the first exact sequence of the Proposition.

Suppose $n = 0$. Then (6.33) reduces to

$$0 \to H_{\mathfrak{m}}^1(\operatorname{Ext}_R^1(I_X, I_X)) \to \operatorname{Hom}_R(I_X, H_{\mathfrak{m}}^1(A)) \to H_{\mathfrak{m}}^0(\operatorname{Ext}_R^2(I_X, I_X)) \to 0. \tag{6.34}$$

Since $\operatorname{depth}_{\mathfrak{m}}(A) = 1$, we get $H_{\mathfrak{m}}^0(\operatorname{Ext}^1(I_X, I_X)) = 0$ because $\operatorname{Ext}^1(I_X, I_X) \cong \operatorname{Hom}_R(I_X, A)$, and we get the second exact sequence by combining (6.34) with

$$H_{\mathfrak{m}}^0(\operatorname{Ext}_R^1(I_X, I_X)) \to \operatorname{Ext}_R^1(I_X, I_X) \to H_*^0(\mathcal{N}) \to H_{\mathfrak{m}}^1(\operatorname{Ext}_R^1(I_X, I_X)) \to 0.$$

To prove the duality, we remark that for each $x \in X$, $(\mathcal{I}_X/\mathcal{I}_X^2 \otimes \omega_X)_x$ is Cohen-Macaulay, cf. [**33**]. Hence $\mathcal{E}xt_{\mathcal{O}_X}^i(\mathcal{I}_X/\mathcal{I}_X^2 \otimes \omega_X, \omega_X) = 0$ for $i > 0$ by local Gorenstein duality (or use Remark 6.7 (ii)), and all isomorphisms of (6.28) hold in this case (with $c = 3$). We therefore get the duality isomorphism in the range $2 \leq i \leq n-2$ by combining the first isomorphism of the proposition with (6.30).

Moreover by Gorenstein duality, we have

$$H_{\mathfrak{m}}^{i+1}(K_A \otimes I_X/I_X^2) \cong \operatorname{Ext}_A^{n-i}(K_A \otimes I_X/I_X^2, K_A)^{\vee} \cong \operatorname{Ext}_A^{n-i}(I_X/I_X^2, A)^{\vee}$$

for $i \geq -1$, where the isomorphism to the right follows from [**32**] and the fact that $X$ is locally Gorenstein. Since $H_*^n(\mathcal{N}_X) \cong \operatorname{Ext}_{\mathcal{O}_X}^n(\mathcal{I}_X/\mathcal{I}_X^2, \mathcal{O}_X)$, we can use (6.25) (with $A$ instead of $R$) to see the exactness of

(6.35)

$$0 \to \operatorname{Ext}_A^n(I_X/I_X^2, A) \to \operatorname{Ext}_{\mathcal{O}_X}^n(\mathcal{I}_X/\mathcal{I}_X^2, \mathcal{O}_X) \to$$

$$\operatorname{Hom}_A(I_X/I_X^2, H_{\mathfrak{m}}^{n+1}(A)) \to \operatorname{Ext}_A^{n+1}(I_X/I_X^2, A) \to 0$$

where the zero to the right follows from reasons of dimension ($H_*^{n+1}(\mathcal{N}_X) = 0$). If $n \geq 1$ (resp. $n = 0$), we conclude by comparing (6.35) with the first ( resp. second) exact sequence of this Proposition. □

Since the duality of Proposition 6.8 shows the vanishing of all $H_{\mathfrak{m}}^i(K_A \otimes I_X)$ for $i \leq n$ provided they vanish for $0 \leq i \leq n/2$ (or for $n/2 \leq i \leq n$), we have

COROLLARY 6.9. *Let $X \subset \mathbb{P}^{n+3}$ be a local Gorenstein ACM subscheme of codimension 3, and let $K_A$ be the canonical module of $A = R/I_X$. If*

$$H_{\mathfrak{m}}^i(K_A \otimes I_X) = 0 \textit{ for } 0 \leq i \leq n/2 \textit{ (or for } n/2 \leq i \leq n)$$

*then*

$$H_{\mathfrak{m}}^i(K_A \otimes I_X) = 0 \textit{ for } 0 \leq i \leq n$$

*i.e. $K_A \otimes_A I_X/I_X^2$ (if non-trivial) is a maximal Cohen-Macaulay module.*

COROLLARY 6.10. *Let $X \subset \mathbb{P}^{n+3}$ be a local complete intersection ACM scheme of codimension 3, and let $K_A$ be the canonical module of $A = R/I_X$. Then, we have $R$-module isomorphisms*

$$H^{i+2}(R, A, A) \cong H^{n+2-i}(R, A, A)^{\vee}(n+4), \textit{ for } 0 \leq i \leq n.$$

PROOF. Since

$$H_{\mathfrak{m}}^i(K_A \otimes I_X)(n+4) \cong H^{i+2}(R, A, A) \text{ for } 0 \leq i \leq n$$

by Remark 6.3 (i) and Proposition 6.8, we conclude by the duality of Proposition 6.8. □

REMARK 6.11. (i) If $n \leq 1$ and $X \subset \mathbb{P}^{n+3}$ is a locally Gorenstein ACM scheme, one may show that the duality of the above Corollary holds.

(ii) From $\operatorname{Ext}_{\mathfrak{m}}^{n+2+i}(I_X, I_X) \cong \operatorname{Ext}_R^i(I_X, H_{\mathfrak{m}}^{n+1}(A))$ for $i = 1$ and 2, the spectral sequence (6.24), with $M_1 = M_2 = I_X$, the isomorphism $H_{\mathfrak{m}}^{n+1}(K_A \otimes_R I_X)(n+4) \cong H_{\mathfrak{m}}^{n+1}(\operatorname{Ext}_R^2(I_X, I_X))$ and Remark 6.7 (iii), we see that an ACM scheme $X \subset \mathbb{P}^{n+3}$ of codimension 3 satisfies

$$H_{\mathfrak{m}}^{n+1}(K_A \otimes_R I_X) \cong \operatorname{Ext}_R^1(I_X, H_{\mathfrak{m}}^{n+1}(A))(-n-4) \cong H_*^0(\mathcal{N}_X)^{\vee}$$

provided $n \geq 1$ (true for $n = 0$ if we replace $H_*^0(\mathcal{N}_X)^{\vee}$ by $\operatorname{Ext}_R^1(I_X, I_X)^{\vee}$) and

$$\operatorname{Ext}_R^2(I_X, H_{\mathfrak{m}}^{n+1}(A)) \cong A^{\vee}(n+4).$$

REMARK 6.12. In codimension 2 we know the $R$-modules $H^1_*(\mathcal{I}_X), \dots, H^n_*(\mathcal{I}_X)$ of $X \subset \mathbb{P}^{n+2}$ are liaison-invariant (up to shifts and duals, cf. [**52**], Theorem 5.3.1) and that they vanish (i.e. $X$ is an ACM subscheme) if and only if $X$ is licci. Moreover, given any set $M_1, \dots, M_n$ of finite length graded $R$-modules, there is a codimension two subscheme $X$ of $\mathbb{P}^{n+2}$ and an integer $d$ such that $H^i_*(\mathcal{I}_X) \cong M_i(d)$ for $1 \le i \le n$ (cf. [**23**]). Furthermore, at least in $\mathbb{P}^3$, we know that $X$ can be taken to be smooth, by [**63**].

In codimension 3 we see that the graded $R$- modules

$$H^0_{\mathfrak{m}}(K_A \otimes_R I_X), \dots, H^n_{\mathfrak{m}}(K_A \otimes_R I_X) \tag{6.36}$$

(recall $H^i_*(\mathcal{N}_X) \cong H^{i-1}_{\mathfrak{m}}(K_A \otimes_R I_X)(n+4)$ for $1 \le i \le n-1$) of an ACM subscheme $X \subset \mathbb{P}^{n+3}$ are liaison-invariant in a degree preserving way and that they therefore vanish if $X$ is licci (Corollary 6.5). Inspired by the codimension 2 results, and being of great interest in itself, we ask:

> Is the converse true, i.e. is an ACM scheme $X$ licci if all the groups of (6.36) vanish?

Unfortunately there are examples of projectively equivalent non-licci curves in $\mathbb{P}^4$ whose modules $H^0_{\mathfrak{m}}(K_A \otimes_R I_X)$ are all isomorphic (they are 3-dimensional, supported in degree 3), but which belong to different CI-liaison classes [**69**]. This indicates that the invariants of (6.36) do not separate CI-liaison classes well. However, to our knowledge our question is still open.

Moreover, in $\mathbb{P}^{n+3}$ for small $n$ (at least say $n \le 6$), one may ask if any set of finite length $R$-modules $N^0, \dots, N^n$ satisfying $N^{n-i}(n+4) \cong (N^i)^\vee$ for $0 \le i \le n$, is realized by some local complete intersection ACM subscheme $X \subset \mathbb{P}^{n+3}$ of codimension 3 in the sense that $N^i \cong H^i_{\mathfrak{m}}(K_A \otimes_R I_X)$, $0 \le i \le n$. The reason why we restrict to small $n$ is due to a famous conjecture of Hartshorne. Indeed if we think the following statement "any local complete intersection ACM subscheme $X$ of $\mathbb{P}^{n+3}$ of dimension $n > 6$ is a complete intersection" is true, then $H^i_{\mathfrak{m}}(K_A \otimes_R I_X)$ must vanish for $0 \le i \le n$ provided $n > 6$; cf. Corollary 6.5.

As an application of Remark 6.11 (ii) and Proposition 6.8 one proves that the usual formula $\chi(\mathcal{N}_C(v)) = 5d + 1 - g + 3dv$ *holds for any generically complete intersection, locally licci, ACM curve* $C$ *in* $\mathbb{P}^4$, cf. Remark 6.4. Indeed the cokernel of $\mathcal{N}_C(-1) \hookrightarrow \mathcal{N}_C$, the multiplication by a general linear form, is supported at $3d$ points, taking care of the $3dv$ term. To see the constant term of $\chi(\mathcal{N}_C(v))$ we apply ${}_v\operatorname{Hom}(-, H^2_{\mathfrak{m}}(A))$ to a minimal free graded resolution

$$\cdots \bigoplus_i R(-n_{2,i}) \to \bigoplus_i R(-n_{1,i}) \to I_C \to 0$$

of $I_C$. We get a complex of terms $\bigoplus_i H^1(\mathcal{O}_C(n_{j,i} + v))$, $1 \le j \le 3$, whose cohomology is ${}_v\operatorname{Ext}^{j-1}_R(I_C, H^2_{\mathfrak{m}}(A))$. For $v \ll 0$ it follows that

$$\sum (-1)^i \dim_v \operatorname{Ext}^i_R(I_C, H^2_{\mathfrak{m}}(A)) = \sum_j (-1)^j \sum_i \chi(\mathcal{O}_C(n_{j,i} + v))$$

which by the Riemann-Roch Theorem is $-(dv + 1 - g)$. Proposition 6.8 shows, however, that

$$-\chi(\mathcal{N}_C(v)) = \dim{}_v \operatorname{Hom}(I_C, H^2_{\mathfrak{m}}(A))$$

for $v \ll 0$, and combining with the information of Remark 6.11 (ii) on the higher ${}_v\operatorname{Ext}^i_R(I_C, H^2_{\mathfrak{m}}(A))$-groups, we get

$$-(dv+1-g) = -\chi(\mathcal{N}_C(v)) - \chi(\mathcal{N}_C(-v-5)) + \chi(\mathcal{O}_C(-v-5))$$

for $v \ll 0$, and we easily find the constant term of $\chi(\mathcal{N}_C(v))$ to be $5d+1-g$, as required. Finally we claim that the argument above extends so that *any generically complete intersection, locally licci and equidimensional curve $C$ in $\mathbb{P}^4$* satisfies $\chi(\mathcal{N}_C(v)) = 5d+1-g+3dv$. The main reason for this is that we only needed $v \ll 0$ in the argument, which allows us to reconsider (6.33) in the proof of Proposition 6.8 and Remark 6.11 (ii) and still get ${}_v\operatorname{Ext}^i_R(I_C, H^2_{\mathfrak{m}}(A))$, for $i = 0$ and $1$ ($v \ll 0$), equal to $-\chi(\mathcal{N}_C(v))$ and $\chi(\mathcal{N}_C(-v-5))$ respectively because ${}_v\operatorname{Ext}^i_R(I_C, H^1_{\mathfrak{m}}(A)) = 0$ for $v \ll 0$. The higher ${}_v\operatorname{Ext}^i_R(I_C, H^2_{\mathfrak{m}}(A))$ groups ($i = 2, 3$) still give $\chi(\mathcal{O}_C(-v-5))$ for $v \ll 0$, and even though the minimal resolution of $I_C$ is one term longer, the alternating sum of these $\operatorname{Ext}^i$-groups is $-(dv+1-g)$ by essentially the same argument. This proves the claim and in particular the following proposition (cf. Remark 6.4):

PROPOSITION 6.13. *Let $C$ be any equidimensional, local complete intersection curve in $\mathbb{P}^4$ of degree $d$, arithmetic genus $g$ and normal sheaf $\mathcal{N}_C$. Then*

$$\chi(\mathcal{N}_C(v)) = 5d+1-g+3dv.$$

EXAMPLE 6.14. To illustrate the results obtained so far, we consider a class of examples of non-licci curves. (In the case $s \geq 3$ or $c_0 = 0$, where the constants are defined below, one may also use [**36**], Corollary 5.13 to see that the curves are not licci.) Let $C$ be a local complete intersection curve in $\mathbb{P}^4$ of degree $d$ and (arithmetic) genus $g$, whose homogeneous ideal $I_C$ has an "almost linear" minimal resolution in the following sense:

$$0 \to R(-s-3)^{\oplus a} \to R(-s-2)^{\oplus b} \to R(-s-1)^{\oplus c_1} \oplus R(-s)^{\oplus c_0} \to I_C \to 0.$$

(So, if $c_0 = 0$, the resolution is linear). We *claim* that $C$ is not licci provided

$$d+g-1-ac_0 \neq 0.$$

Hence the only curve with a linear minimal resolution ($c_0 = 0$) which possibly could be licci, is a straight line (which of course is licci). In particular the smooth rational quartic, which has a linear resolution given by $(a, b, c_1, c_0, s) = (3, 8, 6, 0, 1)$ is not licci, which is well known.

The proof of the claim is nothing more than a computation of the liaison-invariant constant $l(C)_v$, defined by $l(C)_v = \dim\ {}_{v+5}H^0_{\mathfrak{m}}(K_A \otimes_R I_C)$, of Proposition 6.8. Note that the exact sequence and the duality of Proposition 6.8 lead to

$$l(C)_v - l(C)_{-v-5} = h^1(\mathcal{N}_C(v)) - {}_v hom_R(I_C, H^2_{\mathfrak{m}}(A))$$

(small letters refer to dimension). We want to compute the dimensions on the right side for some $v$. Indeed sheafifying the minimal resolution above and taking cohomology, we get

$$H^1(\mathcal{O}_C(s-1)) = 0 \quad \text{and} \quad h^1(\mathcal{O}_C(s-2)) = a.$$

Therefore if we apply ${}_v\operatorname{Hom}_R(-, H^2_{\mathfrak{m}}(A)) \cong {}_v\operatorname{Hom}_R(-, H^1_*(\mathcal{O}_C))$ to the minimal resolution of $I_C$, we get at once

$${}_{-2}hom_R(I_C, H^2_{\mathfrak{m}}(A)) = ac_0 \quad \text{and} \quad {}_{-3}ext^1_R(I_C, H^2_{\mathfrak{m}}(A)) = 0.$$

Observing that

$$h^0(\mathcal{N}_C(-v-5)) = {}_v ext^1_R(I_C, H^2_{\mathfrak{m}}(A))$$

by Remark 6.11 (ii), we get $h^0(\mathcal{N}_C(-2)) = 0$, i.e.

$$h^1(\mathcal{N}_C(-2)) = -\chi(\mathcal{N}_C(-2)) = -(5d+1-g+3d(-2)) = d+g-1,$$

and the claim is proved because

$$l(C)_{-2} - l(C)_{-3} = h^1(\mathcal{N}_C(-2)) - {}_{-2}\, hom_R(I_C, H^2_{\mathfrak{m}}(A)) = d+g-1-ac_0 \neq 0.$$

As a special case, we consider curves $C = C(t)$ with linear resolution

$$0 \to R(-t-2)^{\oplus a} \to R(-t-1)^{\oplus b} \to R(-t)^{\oplus c_1} \to I_C \to 0.$$

Note that determinantal curves defined by a linear matrix of size $t \times (t+2)$, to which we later return, have a minimal resolution of this form. Now, from $H^i(\mathcal{I}_C(v)) = 0$, for $t-2 \leq v \leq t-1$ and $i = 0,1,2$, and Riemann-Roch, we see easily that the degree $d_t$ and the genus $g_t$ of $C$ are

$$d_t = \binom{t+3}{4} - \binom{t+2}{4}, \quad g_t = (t-1)d_t + 1 - \binom{t+3}{4}.$$

Since $d_t + g_t - 1 = td_t - \binom{t+3}{4} > d_{t'} + g_{t'} - 1$ for $t > t'$, it follows that $C(t)$ and $C(t')$ belong to different CI-liaison classes for $t \neq t'$. Varying $t$, we get infinitely many different CI-liaison classes containing determinantal curves; cf. [**69**], Theorem 2.6, for other classes.

REMARK 6.15. We have proved that $H^i_{\mathfrak{m}}(K_A \otimes_R I_X)$ are liaison-invariants and we have used this fact to see that there are infinitely many different CI-liaison classes containing standard determinantal curves in $\mathbb{P}^4$. By the generalization of Gaeta's Theorem, Theorem 3.6, we know that all standard determinantal schemes are glicci. Therefore $H^i_{\mathfrak{m}}(K_A \otimes_R I_X)$ are not G-liaison-invariants.

REMARK 6.16. In the ACM case a main result of [**14**], Theorem 2.3, implies that the groups $H^i_{\mathfrak{m}}(\mathrm{Ext}^j_R(R/I_X, I_X))$, $0 \leq i \leq \dim X$ and $1 \leq j \leq \operatorname{codim} X$, are liaison-invariant. Let us see one easy and interesting case, namely the liaison-invariance of

$$H^0_{\mathfrak{m}}(\mathrm{Ext}^3_R(R/I_X, I_X)) \cong H^0_{\mathfrak{m}}(\mathrm{Ext}^2_R(I_X, I_X)).$$

Note that this group is liaison-invariant in codimension 3 by Proposition 6.8, but we claim its liaison-invariance to hold in any codimension. Indeed we can suppose $\operatorname{codim} X \geq 2$. If $n = \dim X$, we get

$$H^0_{\mathfrak{m}}(\mathrm{Ext}^2_R(I_X, I_X)) \cong H^1_*(X, \mathcal{N}_X) \text{ for } n \geq 2$$

by (6.32), and $H^0_{\mathfrak{m}}(\mathrm{Ext}^2_R(I_X, I_X) \cong ker\delta_X$ for $n = 1$ by (6.33) (with notation as in Theorem 6.1), because the spectral sequence argument used in (6.32), (resp. in (6.33)) holds in codimension $> 3$ as well provided $i = 0$ (resp. $n = 1$). We conclude by Theorem 6.1, a theorem which takes care of the zero-dimensional case as well because, in this case, the exact sequence (6.34) shows that $H^0_{\mathfrak{m}}(\mathrm{Ext}^2_R(I_X, I_X)) \cong \operatorname{coker} \delta_X$. One reason for our interest in the liaison-invariance of this group is the isomorphism

$$H^0_{\mathfrak{m}}(\mathrm{Ext}^2_R(I_X, I_X)) \cong H^0_{\mathfrak{m}}(H^2(R, A, A))$$

which now follows from Remark 6.3 (i). Hence $H^0_{\mathfrak{m}}(H^2(R, A, A))$ is liaison-invariant if $\operatorname{depth}_{\mathfrak{m}} A \geq 1$. Note that, for ACM subschemes of codimension 3, we therefore have in general

$$H^0_{\mathfrak{m}}(K_A \otimes_R I_X) \cong H^0_{\mathfrak{m}}(H^2(R, A, A))(-n-4).$$

Huneke [**35**] (resp. Buchweitz-Ulrich [**14**]) has shown that $H^2(R, A, A) = 0$ provided $A$ is a graded or local Gorenstein ring (resp. generically a complete intersection) which is in the linkage class of a complete intersection (true also when $\dim A = 0$). We can prove

PROPOSITION 6.17. *Let $A = R/I$ be licci as a graded algebra.*

(i) *If $A$ satisfies the condition $G_0$, then $H^2(R, A, A) = 0$.*

(ii) *If $A$ is a complete intersection in codimension $\leq r$ then*

$$H^{i+2}(R, A, A) = 0 \quad \textit{for } 0 \leq i \leq r.$$

PROOF. (i) Since $A$ is generically Gorenstein, one knows that $H^2(R, A, A)_{\mathfrak{q}} = 0$, for all $\mathfrak{q} \in \operatorname{Ass}(A)$ by Huneke's result. If $H^2(R, A, A) \neq 0$, then there exists a graded prime ideal $\mathfrak{p} \subset A$, $\mathfrak{p} \notin \operatorname{Ass}(A)$, such that

$$H^2(R, A, A)_{\mathfrak{p}} \neq 0 \quad \text{and} \quad H^2(R, A, A)_{\mathfrak{p}'} = 0$$

for all graded primes $\mathfrak{p}' \subset \mathfrak{p}$ . We get

$$H^0_{\mathfrak{p}A_{\mathfrak{p}}}(H^2(R, A, A)_{\mathfrak{p}}) \neq 0.$$

By Remark 6.16 we know this group is liaison-invariant. Since $A_{\mathfrak{p}}$ is licci by assumption, we get a contradiction and (i) is proved.

(ii) By induction it suffices to prove $H^{r+2}(R, A, A) = 0$. If $\dim A_{\wp} \leq r$, $\wp$ a graded prime ideal, then $H^{r+2}(R, A, A)_{\wp} = 0$ because $A_{\wp}$ is a complete intersection. And if $\dim A_{\wp} \geq r + 1$, then $H^0_{\wp A_{\wp}}(H^{r+2}(R, A, A)_{\wp})$ is liaison-invariant by Remark 6.3 (i). Since $A$ is licci, this last group vanishes and we are done. □

CHAPTER 7

# Geometric Applications of the CI-Liaison Invariants

In this chapter we consider Cartier divisors $C$ on $S \subset \mathbb{P}^{n+3}$ of dimension $\dim C = n$, where $C = \operatorname{Proj}(A)$ and $S = \operatorname{Proj}(B)$ are ACM subschemes and $S$ satisfies the condition $G_0$. Under some reasonable assumptions we show that there exists a "computable" quotient $L^i(C)$ of $H^i_{\mathfrak{m}}(K_A(n+4) \otimes_R I)$ with several interesting properties. Firstly if $S$ also satisfies $G_1$, then $L^0(C)$ is a quotient of (resp. isomorphic to) $H^2(R, A, A)$ (resp. if $n = 0$), making it possible to find $\dim H^2(R, A, A)$ easily in several situations. Secondly, $L^i(C)$ is invariant under adding hyperplane sections in a non-degree-preserving way, and the assumption $L^i(C) \neq 0$ therefore leads to a criterion which implies that for large values of $t$, $C_t \in |C + tH|$ belong to different CI-liaison classes, in contrast to our earlier results (Proposition 5.12, Corollary 5.14) which show that they always belong to the same Gorenstein liaison class. In Chapter 8, we give a classification of minimal ACM curves , $C$, lying on a smooth rational ACM surface $S \subset \mathbb{P}^4$. Using this classification, we can show that many non arithmetically Gorenstein curves, $C \notin |tH|$, on a cubic scroll, a Castelnuovo surface or a Bordiga surface in $\mathbb{P}^4$ give rise to an infinite number of CI-liaison classes by "adding hyperplane sections" (see Proposition 8.13). For standard determinantal subschemes, which we know belong to the same Gorenstein liaison class (Theorem 3.6), the groups $L^i(C)$ are particularly easy to compute. They are quite often non-vanishing, leading to many different non-licci classes of ACM codimension 3 subschemes in different dimensions $n$, consistent with what Ulrich has proved for some classes of determinantal curves (i.e. $n = 1$; see [**69**], Theorem 2.6). Moreover our result applies to subschemes whose minimal resolutions of their ideals are far from being linear, i.e. it applies to subschemes not covered by [**36**], Corollary 5.13. We will need the following result, whose first part we state for subschemes $C \subset \mathbb{P}^{n+c}$ of codimension $c$ for later use.

LEMMA 7.1. *Let $C \subset S$ be equidimensional, locally Cohen-Macaulay subschemes of $\mathbb{P}^{n+c}$, let $A = R/I_C$, $B = R/I_S$ and suppose $C$ is a Cartier divisor on $S$ of dimension $\dim C = n$.*

(i) *If $S \subset \mathbb{P}^{n+c}$ satisfies $H^2(R, B, B)^\sim = 0$ (e.g. $S$ is locally licci and has the property $G_0$), then $\mathcal{N}_S \otimes \mathcal{O}_C \cong \mathcal{H}om(\mathcal{I}_S, \mathcal{O}_C)$ are isomorphic and there is an exact sequence*

$$0 \to \mathcal{N}_S \otimes \mathcal{I}_{C|S} \to \mathcal{N}_S \to \mathcal{N}_S \otimes \mathcal{O}_C \to 0.$$

(ii) *If $c = 3$ and $S \subset \mathbb{P}^{n+3}$ satisfies $G_1$, then $H^2(R, A, A)^\sim = 0$, i.e.*

$$H^0_{\mathfrak{m}}(H^2(R, A, A)) \cong H^2(R, A, A).$$

PROOF. (i) We claim that $\mathcal{N}_S \otimes \mathcal{O}_C \cong \mathcal{H}om(\mathcal{I}_S, \mathcal{O}_C)$. Indeed the cokernel of

$$\mathcal{N}_S \cong \mathcal{H}om(\mathcal{I}_S, \mathcal{O}_S) \to \mathcal{H}om(\mathcal{I}_S, \mathcal{O}_C)$$

is contained in $H^2(R, B, I_{C|S})^\sim$, with $I_{C|S} = I_C/I_S$, which vanishes by Proposition 6.17 because $\mathcal{I}_{C|S}$ is invertible. Since $\mathcal{N}_S \otimes \mathcal{I}_{C|S} \cong \mathcal{H}om(\mathcal{I}_S, \mathcal{I}_{C|S})$, the exact sequences

$$0 \to \mathcal{H}om(\mathcal{I}_S, \mathcal{I}_{C|S}) \to \mathcal{H}om(\mathcal{I}_S, \mathcal{O}_S) \to \mathcal{H}om(\mathcal{I}_S, \mathcal{O}_C) \to 0$$

and

$$\mathcal{N}_S \otimes \mathcal{I}_{C|S} \to \mathcal{N}_S \to \mathcal{N}_S \otimes \mathcal{O}_C \to 0$$

must coincide, and we conclude easily.

(ii) If $S$ has the property $G_1$, then $H^3(R, B, B)^\sim = 0$ by Proposition 6.17 (ii). Moreover, $H^3(R, B, I_{C|S})^\sim = 0$ since $\mathcal{I}_{C|S}$ is invertible. Combining with $H^2(R, B, B)^\sim = 0$, we get $H^2(R, B, A)^\sim = 0$ and hence $H^2(R, A, A)^\sim = 0$ (because $H^2(B, A, A)^\sim = 0$) by two standard exact sequences of algebra cohomology. We then easily get (ii). $\square$

In what follows, $C$ is a Cartier divisor on $S \subset \mathbb{P}^{n+3}$, and $\dim C = n$, where $C = \operatorname{Proj}(A)$ and $S = \operatorname{Proj}(B)$ are ACM subschemes. Beginning with Proposition 7.2 we will also assume that $S$ has the property $G_0$, but we will not need this assumption before then. We know the homogeneous ideal $I_S$ of $S \subset \mathbb{P}^{n+3}$ has a minimal free resolution

$$0 \to \bigoplus_i R(-q_i) \to \bigoplus_i R(-p_i) \to I_S \to 0. \tag{7.1}$$

Since

$$\begin{aligned} \mathcal{E}xt^1_{\mathcal{O}_{\mathbb{P}}}(\mathcal{I}_S, \mathcal{I}_{C|S}) &\cong \mathcal{E}xt^1_{\mathcal{O}_{\mathbb{P}}}(\mathcal{I}_S, \mathcal{O}_S) \otimes \mathcal{I}_{C|S} \\ &\cong \mathcal{E}xt^1_{\mathcal{O}_{\mathbb{P}}}(\mathcal{I}_S, \mathcal{O}_{\mathbb{P}}) \otimes \mathcal{I}_{C|S} \\ &\cong \omega_S(n+4) \otimes \mathcal{I}_{C|S} \end{aligned}$$

it induces a complex

$$\bigoplus_i H^0(\mathcal{I}_{C|S}(p_i + v)) \to \bigoplus_i H^0(\mathcal{I}_{C|S}(q_i + v)) \to H^0(\omega_S(n+4) \otimes \mathcal{I}_{C|S}(v)) \tag{7.2}$$

for any $v$, which will be important in several chapters. Indeed, in Chapter 9 we shall see that the homology "in the middle" of (7.2) (which is $H^1(\mathcal{N}_S \otimes \mathcal{I}_{C|S}(v))$, cf. below) carries relevant information for deforming $C \subset S \subset \mathbb{P}^{n+3}$ as a flag. Now we show that, under some conditions, the cokernel $L^0(C)_v$ of the right map is a quotient of the liaison-invariant group ${}_vH^0_{\mathfrak{m}}(K_A(n+4) \otimes_R I)$ provided $v \geq$ some $v_0$. If $H$ is a hyperplane section of $S \subset \mathbb{P}^{n+3}$, we obviously have an isomorphism

$$L^0(C)_v \cong L^0(C_t)_{v+t}, \qquad \text{for any } C_t \in |C + tH|$$

(where $C_t$ is given by some non-zero divisor of $H^0(\mathcal{O}_S(C)(t))$) leading to the following interesting fact: if $L^0(C)_v \neq 0$ for some $v \geq v_0$, then $C_t$ is not licci for any $t \geq 0$, and $C$ and $C_t$ ($t \gg 0$) belong to different CI-liaison classes. Moreover if $S$ has $G_1$, we have ${}_vH^0_{\mathfrak{m}}(K_A \otimes_R I)(n+4) \cong {}_vH^2(R, A, A)$ by Remark 6.16 and Lemma 7.1, which shows the direct relevance of these liaison-invariant groups to the deformation theory of $C \subset \mathbb{P}^{n+3}$.

By definition,

$$L^0(C)_v = \operatorname{coker}[\bigoplus_i H^0(\mathcal{I}_{C|S}(q_i+v)) \to H^0(\omega_S(n+4)\otimes \mathcal{I}_{C|S}(v))].$$

We define

$$L^j(C)_v = H^j(\omega_S(n+4)\otimes \mathcal{I}_{C|S}(v)) \quad \text{if } j \geq 1. \tag{7.3}$$

Before proving our results, we will give another description of $L^0(C)_v$. Note that the sequence

$$\bigoplus_i H^0(\mathcal{I}_{C|S}(p_i+v)) \to \bigoplus_i H^0(\mathcal{I}_{C|S}(q_i+v)) \to \operatorname{Ext}^1_{\mathcal{O}_{\mathbb{P}}}(\mathcal{I}_S,\mathcal{I}_{C|S}(v)) \to 0 \tag{7.4}$$

is exact if $n \geq 1$ (replace $\operatorname{Ext}^1_{\mathcal{O}_{\mathbb{P}}}(\mathcal{I}_S,\mathcal{I}_{C|S}(v))$ by ${}_v\operatorname{Ext}^1_R(I_S,I_{C|S})$ if $n=0$) and that the spectral sequence $H^p(\mathcal{E}xt^q_{\mathcal{O}_{\mathbb{P}}}(\mathcal{I}_S,\mathcal{I}_{C|S}(v)))$ implies the exactness of

(7.5)
$$0 \to H^1(\mathcal{N}_S\otimes\mathcal{I}_{C|S}(v)) \to \operatorname{Ext}^1_{\mathcal{O}_{\mathbb{P}}}(\mathcal{I}_S,\mathcal{I}_{C|S}(v)) \to H^0(\omega_S(n+4)\otimes\mathcal{I}_{C|S}(v))$$
$$\to H^2(\mathcal{N}_S\otimes\mathcal{I}_{C|S}(v)) \to \cdots$$

because $\mathcal{E}xt^1_{\mathcal{O}_{\mathbb{P}}}(\mathcal{I}_S,\mathcal{I}_{C|S}) \cong \omega_S(n+4)\otimes\mathcal{I}_{C|S}$. It follows that the homology group "in the middle" of (7.2) is $H^1(\mathcal{N}_S\otimes\mathcal{I}_{C|S}(v))$ and that

$$L^0(C)_v = \ker[H^2(\mathcal{N}_S\otimes\mathcal{I}_{C|S}(v)) \to \operatorname{Ext}^2_{\mathcal{O}_{\mathbb{P}}}(\mathcal{I}_S,\mathcal{I}_{C|S}(v))] \tag{7.6}$$

provided $n \geq 1$ (for $n > 1$, the $\operatorname{Ext}^2$-group vanishes, cf. the proof below). To prove that $L^0(C)_v$ is a certain quotient of ${}_vH^0_{\mathfrak{m}}(K_A\otimes_R I)$, we treat the seemingly most difficult case $n \leq 1$ in the following proposition, while the case $n \geq 2$ is considered in Proposition 7.3.

PROPOSITION 7.2. *Let $C \subset S \subset \mathbb{P}^{n+3}$ be two ACM subschemes, $S$ generically a complete intersection in $\mathbb{P}^{n+3}$, and suppose $C$ is a Cartier divisor on $S$ of dimension $n = \dim C$. Let $A = R/I$, $I = I_C$, and suppose, for some $v$, that*

$${}_v\operatorname{Hom}_R(I, H^{n+1}_{\mathfrak{m}}(A)) \to {}_v\operatorname{Hom}_R(I_S, H^{n+1}_{\mathfrak{m}}(A))$$

*is injective (with cokernel $T_v$). Then the natural map $H^n(\mathcal{N}_{C|S}(v)) \to H^n(\mathcal{N}_C(v))$ (induced from the short exact sequence $0 \to \mathcal{N}_{C|S} \to \mathcal{N}_C \to \mathcal{N}_S\otimes\mathcal{O}_C \to 0$) factors via ${}_vH^{n-1}_{\mathfrak{m}}(K_A(n+4)\otimes_R I) \cong {}_vH^{n-1}_{\mathfrak{m}}(\operatorname{Ext}^2_R(I,I)) \hookrightarrow H^n(\mathcal{N}_C(v))$, and we have*

(i) *for $n \geq 1$, two exact sequences*

$$H^n(\mathcal{N}_{C|S}(v)) \to {}_vH^{n-1}_{\mathfrak{m}}(K_A(n+4)\otimes_R I) \to L^{n-1}(C)_v \to 0$$

$$0 \to {}_vH^n_{\mathfrak{m}}(K_A(n+4)\otimes_R I) \to L^n(C)_v \to T_v \to 0$$

(ii) *for $n = 0$, the isomorphisms*

$${}_vH^0_{\mathfrak{m}}(K_A(4)\otimes_R I) \xrightarrow{\sim} {}_vH^1_{\mathfrak{m}}(K_B(4)\otimes_B I_{C|S}) \cong L^0(C)_v \qquad \textit{provided } T_v = 0$$

*where $K_B$ is the canonical module $\operatorname{Ext}^2_R(B,R)(-4) \cong H^0_*(\omega_S)$ of $B = R/I_S$.*

PROOF. For $n \geq 1$ we *claim* that $H^n(\mathcal{N}_{C|S}(v)) \to H^n(\mathcal{N}_C(v))$ factors via

$${}_vH^{n-1}_{\mathfrak{m}}(K_A(n+4)\otimes_R I) \hookrightarrow H^n(\mathcal{N}_C(v))$$

and that the cokernel of

$$H^n(\mathcal{N}_{C|S}(v)) \to {}_vH^{n-1}_{\mathfrak{m}}(K_A(n+4)\otimes_R I) \cong {}_vH^{n-1}_{\mathfrak{m}}(\operatorname{Ext}^2_R(I,I))$$

is isomorphic to $\ker \phi_{C|S}$, where $\phi_{C|S} : H^n(\mathcal{N}_S \otimes \mathcal{O}_C(v)) \to {}_v \operatorname{Hom}_R(I_S, H^{n+1}_{\mathfrak{m}}(A))$. Indeed we have $\mathcal{N}_S \otimes \mathcal{O}_C \cong \mathcal{H}om(\mathcal{I}_S, \mathcal{O}_C)$ by Lemma 7.1, which leads to an exact sequence

$$0 \to \mathcal{N}_{C|S} \to \mathcal{N}_C \to \mathcal{N}_S \otimes \mathcal{O}_C \to 0.$$

We therefore get a commutative diagram of exact horizontal sequences;

(7.7)

$$\begin{array}{ccccccc} H^n(\mathcal{N}_{C|S}(v)) & \to & H^n(\mathcal{N}_C(v)) & \to & H^n(\mathcal{N}_S \otimes \mathcal{O}_C(v)) & \to & 0 \\ & & \downarrow \delta_C & \circ & \downarrow \phi_{C|S} & & \\ 0 & \to & {}_v \operatorname{Hom}_R(I, H^{n+1}_{\mathfrak{m}}(A)) & \to & {}_v \operatorname{Hom}_R(I_S, H^{n+1}_{\mathfrak{m}}(A)) & \to & \end{array}$$

and we get the claim because ${}_vH^{n-1}_{\mathfrak{m}}(K_A(n+4) \otimes_R I) \cong \ker \delta_C$ by Proposition 6.8. Since ${}_vH^n_{\mathfrak{m}}(K_A(n+4) \otimes_R I) \cong \operatorname{coker} \delta_C$ (true also when $n = 0$), the diagram above gives the exact sequence

$$0 \to {}_vH^n_{\mathfrak{m}}(K_A(n+4) \otimes_R I) \to \operatorname{coker} \phi_{C|S} \to T_v \to 0. \tag{7.8}$$

Next we claim that $\phi_{C|S}$ and

$$\rho_{C|S} : H^{n+1}(\mathcal{N}_S \otimes \mathcal{I}_{C|S}(v)) \to \operatorname{Ext}^{n+1}_{\mathcal{O}_{\mathbb{P}}}(\mathcal{I}_S, \mathcal{I}_{C|S}(v))$$

have isomorphic kernels and cokernels (for $n \geq 1$). Indeed if we consider the exact sequence (6.25), i.e.

$$\to {}_v \operatorname{Ext}^{n+1}_R(I_S, I_{C|S}) \to \operatorname{Ext}^{n+1}_{\mathcal{O}_{\mathbb{P}}}(\mathcal{I}_S, \mathcal{I}_{C|S}(v)) \to {}_v \operatorname{Ext}^{n+2}_{\mathfrak{m}}(I_S, I_{C|S}) \to$$

and if we use $\operatorname{pd}_B I_S = 1$, we get

$$\operatorname{Ext}^{n+1}_{\mathcal{O}_{\mathbb{P}}}(\mathcal{I}_S, \mathcal{I}_{C|S}(v)) \cong {}_v \operatorname{Ext}^{n+2}_{\mathfrak{m}}(I_S, I_{C|S}) \cong {}_v \operatorname{Hom}_R(I_S, H^{n+2}_{\mathfrak{m}}(I_{C|S}))$$

by (6.23). Arguing similarly and using $\mathcal{N}_S \cong \mathcal{E}xt^1_{\mathcal{O}_{\mathbb{P}}}(\mathcal{I}_S, \mathcal{I}_S)$, we get

$$\operatorname{Ext}^{i+1}_{\mathcal{O}_{\mathbb{P}}}(\mathcal{I}_S, \mathcal{I}_{C|S}(v)) = 0 \quad \text{for } 0 < i < n$$

$$H^n(\mathcal{N}_S(v)) \cong \operatorname{Ext}^{n+1}_{\mathcal{O}_{\mathbb{P}}}(\mathcal{I}_S, \mathcal{I}_S(v)) = 0$$

$$H^{n+1}(\mathcal{N}_S(v)) \cong \operatorname{Ext}^{n+2}_{\mathcal{O}_{\mathbb{P}}}(\mathcal{I}_S, \mathcal{I}_S(v)) \cong {}_v \operatorname{Hom}_R(I_S, H^{n+3}_{\mathfrak{m}}(I_S)).$$

Now we consider the following diagram

$$\begin{array}{ccc} H^n(\mathcal{N}_S(v)) & & 0 \\ \downarrow & & \downarrow \\ H^n(\mathcal{N}_S \otimes \mathcal{O}_C(v)) & \xrightarrow{\phi_{C|S}} & {}_v \operatorname{Hom}_B(I_S, H^{n+2}_{\mathfrak{m}}(I)) \\ \downarrow & & \downarrow \\ H^{n+1}(\mathcal{N}_S \otimes \mathcal{I}_{C|S}(v)) & \xrightarrow{\rho_{C|S}} & {}_v \operatorname{Hom}_R(I_S, H^{n+2}_{\mathfrak{m}}(I_{C|S})) \\ \downarrow & \circ & \downarrow \\ H^{n+1}(\mathcal{N}_S(v)) & \xrightarrow{\sim} & {}_v \operatorname{Hom}_R(I_S, H^{n+3}_{\mathfrak{m}}(I_S)) \\ \downarrow & & \downarrow \\ 0 & & \end{array}$$

and we get the claim.

In the case $n \geq 1$, we can conclude using the spectral sequence which led to (7.5). Recalling

$$\mathcal{N}_S \otimes \mathcal{I}_{C|S} \cong \mathcal{H}om(\mathcal{I}_S, \mathcal{I}_{C|S}), \qquad \mathcal{E}xt^1_{\mathcal{O}_\mathbb{P}}(\mathcal{I}_S, \mathcal{I}_{C|S}) \cong \omega_S(n+4) \otimes \mathcal{I}_{C|S},$$

$$\text{and} \quad \operatorname{Ext}^{i+1}_{\mathcal{O}_\mathbb{P}}(\mathcal{I}_S, \mathcal{I}_{C|S}(v)) = 0 \ \text{ for } 0 < i < n,$$

the spectral sequence yields the exact sequence

$$H^{n-1}(\mathcal{E}xt^1_{\mathcal{O}_\mathbb{P}}(\mathcal{I}_S, \mathcal{I}_{C|S}(v))) \to H^{n+1}(\mathcal{N}_S \otimes \mathcal{I}_{C|S}(v)) \to \operatorname{Ext}^{n+1}_{\mathcal{O}_\mathbb{P}}(\mathcal{I}_S, \mathcal{I}_{C|S}(v)) \to$$

$$H^n(\mathcal{E}xt^1_{\mathcal{O}_\mathbb{P}}(\mathcal{I}_S, \mathcal{I}_{C|S}(v))) \to 0.$$

Hence, $\operatorname{coker} \rho_{C|S} \cong H^n(\omega_S(n+4) \otimes \mathcal{I}_{C|S}(v)) = L^n(C)_v$ and $\ker \rho_{C|S} \cong L^{n-1}(C)_v$ (the formula for $\ker \rho_{C|S}$ holds also for $n = 1$ by (7.6)) and the proof is complete when $n \geq 1$.

To treat the case $n = 0$, we claim there is an exact sequence

$$0 \to \mathcal{N}_S \otimes \mathcal{O}_C \to \mathcal{E}xt^1_{\mathcal{O}_\mathbb{P}}(\mathcal{I}_S, \mathcal{I}_C) \to \omega_S(4) \otimes \mathcal{I}_{C|S} \to 0$$

and $\mathcal{H}om_{\mathcal{O}_\mathbb{P}}(\mathcal{I}_S, \mathcal{I}_C) \cong \mathcal{O}_\mathbb{P}$. Indeed we prove it using $\operatorname{pd} \mathcal{I}_S = 1$, the isomorphism $\mathcal{N}_S \otimes \mathcal{O}_C \cong \mathcal{H}om_{\mathcal{O}_\mathbb{P}}(\mathcal{I}_S, \mathcal{O}_C)$, the injectivity of

$$\mathcal{E}xt^1(\mathcal{I}_S, \mathcal{I}_{C|S}) \cong \omega_S(4) \otimes \mathcal{I}_{C|S} \to \mathcal{E}xt^1(\mathcal{I}_S, \mathcal{O}_S) \cong \omega_S(4)$$

deduced from Lemma 7.1 (i), and

$$0 \to \mathcal{H}om(\mathcal{I}_S, \mathcal{I}_C) \to \mathcal{H}om(\mathcal{I}_S, \mathcal{O}_\mathbb{P}) \to \mathcal{H}om(\mathcal{I}_S, \mathcal{O}_C) \to \mathcal{E}xt^1(\mathcal{I}_S, \mathcal{I}_C)$$

$$\to \mathcal{E}xt^1(\mathcal{I}_S, \mathcal{O}_\mathbb{P}) \to \mathcal{E}xt^1(\mathcal{I}_S, \mathcal{O}_C) \to 0.$$

Next using the proven claim we get the following diagram of exact sequences

$$\begin{array}{ccccccccc}
 & & & & 0 & & & & \\
 & & & & \downarrow & & & & \\
 & & & & H^0(\mathcal{N}_S \otimes \mathcal{O}_C(v)) & & & & \\
 & & & & \downarrow & \searrow \phi_{C|S} & & & \\
0 & \to & {}_v\operatorname{Ext}^1_R(I_S, I) & \to & H^0(\mathcal{E}xt^1_{\mathcal{O}_\mathbb{P}}(\mathcal{I}_S, \mathcal{I}_C(v))) & \longrightarrow & {}_v\operatorname{Hom}_R(I_S, H^2_\mathfrak{m}(I)) & \to & 0 \\
 & & & & \downarrow & & & & \\
 & & & & H^0(\omega_S(4) \otimes \mathcal{I}_{C|S}(v)) & & & & \\
 & & & & \downarrow & & & & \\
 & & & & 0 & & & &
\end{array}$$

where the horizontal sequence is induced by (6.25), recalling $\operatorname{Ext}^1_{\mathcal{O}_\mathbb{P}}(\mathcal{I}_S, \mathcal{I}_C(v)) \cong H^0(\mathcal{E}xt^1_{\mathcal{O}_\mathbb{P}}(\mathcal{I}_S, \mathcal{I}_C(v)))$. There is then an induced map

$${}_v\operatorname{Ext}^1_R(I_S, I) \to H^0(\omega_S(4) \otimes \mathcal{I}_{C|S}(v))$$

whose cokernel is isomorphic to $\operatorname{coker} \phi_{C|S}$, which by (7.8) and the assumption $T_v = 0$ is ${}_vH^0_\mathfrak{m}(K_A(4) \otimes_R I)$. Since ${}_v\operatorname{Ext}^1_R(I_S, I) \to {}_v\operatorname{Ext}^1_R(I_S, I_{C|S})$ is surjective, the cokernel above is isomorphic to the cokernel of

$${}_v\operatorname{Ext}^1_R(I_S, I_{C|S}) \to H^0(\omega_S(4) \otimes \mathcal{I}_{C|S}(v)) \cong H^0(\operatorname{Ext}^1_R(I_S, I_{C|S})^{\sim}(v))$$

which is ${}_vH^1_\mathfrak{m}(K_B(4) \otimes_B I_{C|S})$, and we get the last isomorphism of (ii). Finally since $\bigoplus_i H^0(\mathcal{I}_{C|S}(q_i + v)) \to {}_v\operatorname{Ext}^1_R(I_S, I_{C|S})$ is surjective, this cokernel is also isomorphic to the cokernel described immediately before (7.3), i.e. to $L^0(C)_v$, and we are done. □

For $n \geq 2$, we have a similar result, replacing the injectivity assumption of Proposition 7.2 by the vanishing of $H^{i+1}(\mathcal{N}_{C|S}(v))$. Note that since we have $\mathcal{N}_{C|S} \cong \omega_C \otimes \omega_S^{-1}$, the vanishing assumption above is sometimes easily seen by reasons of its degree (or the degree of its dual) or by ampleness.

PROPOSITION 7.3. *Let $C \subset S \subset \mathbb{P}^{n+3}$ be two ACM subschemes, $S$ generically a complete intersection, and suppose $C$ is a Cartier divisor on $S$ of dimension $n = \dim C \geq 2$. Let $A = R/I$, $I = I_C$, and let $i$ be an integer, $2 \leq i+1 \leq n$. Then*

$$L^{i-1}(C)_v \cong H^{i+1}(\mathcal{N}_S \otimes \mathcal{I}_{C|S}(v)).$$

*Moreover, if $H^{i+1}(\mathcal{N}_{C|S}(v)) = 0$ (resp. $H^j(\mathcal{N}_{C|S}(v)) = 0$ for $j = i$ and $i+1$) for some $v$, then the composition*

$${}_vH_{\mathfrak{m}}^{i-1}(K_A(n+4) \otimes_R I) \cong H^i(\mathcal{N}_C(v)) \to H^{i+1}(\mathcal{N}_S \otimes \mathcal{I}_{C|S}(v)) \cong L^{i-1}(C)_v$$

*is surjective (resp. bijective).*

PROOF. By the spectral sequence of (7.5), we get

$$H^{i+1}(\mathcal{N}_S \otimes \mathcal{I}_{C|S}(v)) \cong H^{i-1}(\omega_S(n+4) \otimes \mathcal{I}_{C|S}(v))$$

if $i+1 > 2$ (resp. (7.5) if $i+1 = 2$). Combining with (7.3) (resp. (7.6) and remark that $\operatorname{Ext}^2_{\mathcal{O}_{\mathbb{P}}}(\mathcal{I}_S, \mathcal{I}_{C|S}(v)) = 0$ because $n \geq 2$), we get the first isomorphism of the proposition.

By Corollary 6.5, we have $H^i_*(\mathcal{N}_S) = H^{i+1}_*(\mathcal{N}_S) = 0$. The sequence

$$0 \to \mathcal{N}_{C|S} \to \mathcal{N}_C \to \mathcal{N}_S \otimes \mathcal{O}_C \to 0$$

and the exact sequence of Lemma 7.1 yield

$$H^i(\mathcal{N}_C(v)) \twoheadrightarrow H^i(\mathcal{N}_S \otimes \mathcal{O}_C(v)) \cong H^{i+1}(\mathcal{N}_S \otimes \mathcal{I}_{C|S}(v)) \tag{7.9}$$

where we have used the assumption on $H^{i+1}(\mathcal{N}_{C|S}(v))$ to see the surjection. By Proposition 6.8, we have $H_{\mathfrak{m}}^{i-1}(K_A(n+4) \otimes_R I) \cong H^i_*(\mathcal{N}_C)$, so we get the surjection of the proposition.

Finally, if in addition $H^i(\mathcal{N}_{C|S}(v)) = 0$, then (7.9) is an isomorphism, and we get the bijection of the Proposition. □

We want to state the vanishing of ${}_vH_{\mathfrak{m}}^0(K_A(n+4) \otimes_R I) \cong_v H_{\mathfrak{m}}^0(H^2(R, A, A))$ under as weak conditions as possible by the approach we have developed. If $n = 0$, the statement is really given in Proposition 7.2 (ii), but for the case $n \geq 1$ Proposition 7.2 (i) only gives an inequality of dimensions. We can skip the injectivity assumption of Proposition 7.2 as well. Indeed, we have

COROLLARY 7.4. *Let $C \subset S \subset \mathbb{P}^{n+3}$ be two ACM subschemes, $S$ having the Property $G_1$, and suppose $C$ is a Cartier divisor on $S$ of dimension $n = \dim C \geq 1$. Let $A = R/I$, $I = I_C$ and suppose, for some $v$, that $L^0(C)_v = 0$ and that the connecting morphism*

$$H^0(\mathcal{N}_S \otimes \mathcal{O}_C(v)) \to H^1(\mathcal{N}_{C|S}(v))$$

*is surjective. Then*

$${}_vH^2(R, A, A) = 0.$$

PROOF. If $n = 1$, we will use Proposition 7.2 (i); more precisely we will consider its proof, because we have dropped an assumption of Proposition 7.2: in the commutative diagram (7.7) for $n = 1$, the map of the Hom-groups is not necessarily an injection. However, by the surjectivity assumption of this corollary, the upper right map of (7.7), $H^1(\mathcal{N}_C(v)) \to H^1(\mathcal{N}_S \otimes \mathcal{O}_C(v))$, is an isomorphism. Since the proof (of the second claim) of Proposition 7.2 shows

$$\ker \phi_{C|S} = \ker \rho_{C|S} \cong L^0(C)_v$$

which vanishes by assumption, we get the injectivity of $\phi_{C|S}$, and hence the injectivity of $\delta_C$ by using (7.7). By Proposition 6.8,

$${}_vH^0_{\mathfrak{m}}(K_A(n+4) \otimes_R I) = 0.$$

Since Lemma 7.1 (ii) shows ${}_vH^2(R, A, A) \cong {}_vH^0_{\mathfrak{m}}(H^2(R, A, A))$, we conclude by the final statement of Remark 6.16. If $n \geq 2$, we use Proposition 7.3 (i) for $i = 1$. Indeed, since the surjection of Proposition 7.3 is exactly given by (7.9), the surjection fits into the sequence

$$H^0(\mathcal{N}_S \otimes \mathcal{O}_C(v)) \to H^1(\mathcal{N}_{C|S}(v)) \to {}_vH^0_{\mathfrak{m}}(K_A(n+4) \otimes_R I) \to L^0(C)_v \to 0$$

and we conclude by the arguments for the case $n = 1$. □

For subschemes $C \subset S$ in $\mathbb{P}^{n+3}$ for which $C$ is a section of a twist of the canonical divisor on $S$, the computations of the quotient $L^0(C)_v$ of the liaison-invariant group ${}_vH^0_{\mathfrak{m}}(K_A(n+4) \otimes_R I_C) \cong {}_vH^0_{\mathfrak{m}}(H^2(R, A, A))$ become particularly easy when we use the complex (7.2). A special case of such twisted canonical sections is the standard determinantal schemes we studied in Chapter 3, and we will illustrate the power of Proposition 7.2 and Proposition 7.3 by computing $L^0(C)_v$ for a class of examples. (The same example was used by Charles Walter in a talk at the Emile Borel Center, Paris, May 1995, to show that the dimension of the Hilbert scheme of curves in $\mathbb{P}^4$ of degree $d$ and genus $g$ does not allow a linear function in $d$ as a lower bound).

EXAMPLE 7.5. Let $C = \text{Proj}(A) \subset \mathbb{P}^{n+3}$ be a good determinantal subscheme defined by a $t \times (t+2)$ matrix $[L, M]$ where $M$ is a column of entries of degree $m$ and $L$ is a $t \times (t+1)$ matrix of linear entries whose minors define a subscheme $S = \text{Proj}(B)$ of $\mathbb{P}^{n+3}$ containing $C$. To apply Proposition 7.2 and Proposition 7.3, we suppose that $C$ is a Cartier divisor on $S$ of dimension $n \leq 2$ (to make the Cartier assumption possible) and that $S$ satisfies $G_1$. The minimal resolutions of $A$ and $B$ are found by using the exactness of the Eagon-Northcott complex, which gives

$$0 \to R(-t-1)^{\oplus t} \to R(-t)^{\oplus(t+1)} \to I_S \to 0,$$

and

$$0 \to R(-m-t-1)^{\oplus \frac{1}{2}t(t+1)} \to R(-t-1)^{\oplus t} \oplus R(-m-t)^{\oplus t(t+1)} \to$$
$$R(-t)^{\oplus(t+1)} \oplus R(-m-t+1)^{\oplus \frac{1}{2}t(t+1)} \to I_C \to 0.$$

If we consider the diagram (3.1) in the proof of the generalized Gaeta Theorem (Theorem 3.6), one may see that $\mathcal{O}_S(C) \cong \omega_S(n+3-t+m)$ and in particular that the complex (7.2) becomes

$$H^0(\mathcal{I}_{C|S}(t+v))^{\oplus(t+1)} \to H^0(\mathcal{I}_{C|S}(t+v+1))^{\oplus t} \to H^0(\mathcal{O}_S(t-m+1+v)).$$

From the resolution of $I_C$ we see that $H^0(\mathcal{I}_{C|S}(v)) = 0$ for $v \le t+m-2$, and we get

$$L^0(C)_v \cong H^0(\mathcal{O}_S(t-m+1+v)) \quad \text{for } v \le m-3.$$

The resolution of $I_C$ also implies that $H^{n+1}(\mathcal{I}_C(v)) = 0$ for $v > m+t-n-3$. The injectivity assumption of Proposition 7.2 is therefore satisfied (and $T_v = 0$) for $v \ge -n-1$. In the case $n = 1$ we get

$$\dim {}_vH^0_{\mathfrak{m}}(K_A(5) \otimes_R I) \ge \dim L^0(C)_v$$

and by duality

$$_{-v}H^0_{\mathfrak{m}}(K_A(n+4) \otimes_R I)^\vee \cong {}_{v+5}H^1_{\mathfrak{m}}(K_A(n+4) \otimes_R I) \cong L^1(C)_v = 0 \quad \text{for } v \ge -2.$$

To compute $L^0(C)_v$ when $n = 2$ , we see by Proposition 7.3 that we need to prove the vanishing of $H^2(\mathcal{N}_{C|S}(v))$. Since $\mathcal{O}_S(C) \cong \omega_S(n+3-t+m)$ and

$$0 \to \mathcal{O}_S \to \mathcal{O}_S(C) \to \mathcal{N}_{C|S} \to 0$$

is exact, we get the vanishing of $H^i(\mathcal{N}_{C|S}(v))$, $i = 1, 2$, for any $v \ge t-4$. Summing up we have

for $n = 0$:

$$\begin{aligned}\dim {}_vH^0_{\mathfrak{m}}(K_A(4) \otimes_R I) &= \dim L^0(C)_v \\ &= h^0(\mathcal{O}_S(t-m+1+v)), \quad -1 \le v \le m-3;\end{aligned}$$

for $n = 1$:

$$\begin{aligned}\dim {}_vH^0_{\mathfrak{m}}(K_A(5) \otimes_R I) &\ge \dim L^0(C)_v \\ &= h^0(\mathcal{O}_S(t-m+1+v)), \quad -2 \le v \le m-3;\end{aligned}$$

$$\text{and } {}_vH^0_{\mathfrak{m}}(K_A(5) \otimes_R I) = 0 \quad \text{for } v \le -3;$$

for $n = 2$:

$$\begin{aligned}\dim {}_vH^0_{\mathfrak{m}}(K_A(6) \otimes_R I) &= \dim L^0(C)_v \\ &= h^0(\mathcal{O}_S(t-m+1+v)), \quad t-4 \le v \le m-3.\end{aligned}$$

Of course this gives infinitely many non-licci ACM-subschemes, belonging to different CI-liaison classes (also in the case $n = 1$ because the inequality is an equality for large $v$ and $m$: cf. Corollary 7.6 and observe that the adding a hyperplane section amounts to increasing $m$ by 1).

If $n = 1$ we consider the case $v = 0$ and the group ${}_0H^2(R, A, A)$ which plays an important role in deformation theory. Since we suppose $S$ satisfies $G_1$, then we have ${}_vH^2(R, A, A) \cong {}_vH^0_{\mathfrak{m}}(K_A(n+4) \otimes_R I)$ by Remark 6.16 and Lemma 7.1. If $3 \le m \le t+1$, the computations above show that

$$\dim {}_0H^2(R, A, A) \ge h^0(\mathcal{O}_S(t-m+1)) = \binom{t-m+5}{4} \ne 0.$$

Invoking a later result of Chapter 9 (i.e. the vanishing of a certain map $\alpha_L$ in the proof of Proposition 10.7), we can, for $m \ge 3$, show

$$\begin{aligned}\dim {}_0H^2(R, A, A) &= h^1(\mathcal{N}_{C|S}) + h^0(\mathcal{O}_S(t-m+1)) \\ &= h^2(\mathcal{O}_S) - h^0(\mathcal{O}_S(t-m-4)) + h^0(\mathcal{O}_S(t-m+1)).\end{aligned}$$

Studying the terms of the complex (7.2) more carefully in the case $m = 1$ and 2, we get

$$\dim {}_0H^2(R, A, A) \ge h^0(\mathcal{O}_{\mathbb{P}}(t-1)) - t\binom{t+1}{2} = \binom{t+1}{3}(t-6)/4 \ne 0$$

for $t > 6$ and $m = 2$, and moreover

$$\dim {}_0H^2(R, A, A) \geq \binom{t+1}{3}(t-26)/4 \neq 0$$

for $t > 26$ and $m = 1$.

The case $n = 0$ is perhaps even more interesting because if $S$ has $G_1$ and $m \geq 3$ we obtain the equalities

$$\dim {}_0H^2(R, A, A) = h^0(\mathcal{O}_S(t-m+1)) = \begin{cases} \binom{t-m+4}{3} & \text{for } m \leq t+1; \\ 0 & \text{for } m \geq t+2. \end{cases}$$

As mentioned above, adding a hyperplane section to the scheme $C$ of Example 7.5 amounts to increasing $m$ by 1. Propositions 7.2 and 7.3 allow us to see rather concretely how the invariant ${}_vH^0_{\mathfrak{m}}(K_A(n+4) \otimes_R I)$ changes when we add hyperplane sections to a subscheme $C$ of $S \subset \mathbb{P}^{n+3}$. In particular we have

COROLLARY 7.6. *Let $C \subset S \subset \mathbb{P}^{n+3}$ be two ACM subschemes, $S$ generically a complete intersection in $\mathbb{P}^{n+3}$, and suppose $C$ is a Cartier divisor on $S$ of dimension $n = \dim C \geq 1$. Suppose $L^{n-1}(C)_{v_0} \neq 0$ for some $v_0$ and let $C_v \in |C + vH|$ be effective Cartier divisors. Then $C_v$ and $C_{v'}$ are not licci and they belong to different CI-liaison classes for any $v > v' \gg 0$.*

*Indeed, if*

$$\begin{aligned} t &= \max\{\ v \mid {}_v\mathrm{Hom}_R(I_{C|S}, H^{n+2}_{\mathfrak{m}}(I_{C|S})) \neq 0\ \} + 1, \\ a &= \max\{\ v \mid H^{n+1}(\mathcal{O}_S(v)) \neq 0\ \} + 1, \ \textit{and} \\ b &= \max\{\ v \mid H^{n}(\mathcal{O}_S(C)(v)) \neq 0\ \} + 1 \end{aligned}$$

*(put $a, b$ or $t = -\infty$ if the corresponding group vanishes for any $v$), then*

(i) *$C_v$ is not licci for $v \geq t - v_0$, and ${}_{v+v_0}H^{n-1}_{\mathfrak{m}}(K_{A_v}(n+4) \otimes_R I_v) \neq 0$ where $I_v = I_{C_v}$ and $A_v = R/I_v$;*

(ii) *$C_v$ and $C_{v'}$ belong to different CI-liaison classes for any $v$ and $v'$ satisfying*

$$v > v' \geq t - v_0, \ \ v \geq a - v_0 \ \textit{and} \ v + v' \geq b - v_0.$$

*In fact ${}_{u+v_0}H^{n-1}_{\mathfrak{m}}(K_{A_u}(n+4) \otimes_R I_u) \cong L^{n-1}(C)_{v_0}$ for $u \geq v - 1$.*

PROOF. By taking cohomology of $\mathcal{I}_{C_v|S} \cong \mathcal{O}_S(-C_v) \cong \mathcal{I}_{C|S}(-v)$, we see that $I_{C_v|S} \cong I_{C|S}(-v)$ and $H^{n+2}_{\mathfrak{m}}(I_{C_v|S}) \cong H^{n+2}_{\mathfrak{m}}(I_{C|S})(-v)$. Moreover, $H^{n+2}_{\mathfrak{m}}(I_v) \hookrightarrow H^{n+2}_{\mathfrak{m}}(I_{C_v|S})$ and we get

$$\begin{aligned} {}_\mu\mathrm{Hom}_R(I_{C_v|S}, H^{n+2}_{\mathfrak{m}}(I_v)) &\hookrightarrow {}_\mu\mathrm{Hom}_R(I_{C_v|S}, H^{n+2}_{\mathfrak{m}}(I_{C_v|S})) \\ &\cong {}_\mu\mathrm{Hom}_R(I_{C|S}, H^{n+2}_{\mathfrak{m}}(I_{C|S})) \\ &= 0 \end{aligned}$$

for $\mu \geq t$ and for all $v$. Hence ${}_\mu H^{n-1}_{\mathfrak{m}}(K_{A_v}(n+4) \otimes_R I_v) \to L^{n-1}(C_v)_\mu$ is surjective for $\mu \geq t$ and for all $v$ by Proposition 7.2. Since we obviously have

$$L^{n-1}(C_v)_{v+v_0} \cong L^{n-1}(C)_{v_0} \neq 0$$

(choose the largest possible $v_0$ such that the group is non-vanishing) and since $H^{n-1}_{\mathfrak{m}}(K_{A_v}(n+4) \otimes_R I_v)$ is liaison-invariant, it follows that $C_v$ is not licci for $v + v_0 \geq t$, and (i) holds.

If $v > v' \geq t - v_0$, $v \geq a - v_0$ and $v + v' \geq b - v_0$, then $L^{n-1}(C)_{v-v'+v_0} = 0$ because $v - v' + v_0 > v_0$. Then Proposition 7.2 implies

$${}_{v+v_0}H^{n-1}_{\mathfrak{m}}(K_{A_{v'}}(n+4) \otimes_R I_{v'}) \cong L^{n-1}(C_{v'})_{v+v_0} \cong L^{n-1}(C)_{v-v'+v_0} = 0$$

provided we can show $H^n(\mathcal{N}_{C_{v'}|S}(v+v_0)) = 0$. This, however, follows by taking cohomology of

$$0 \to \mathcal{O}_S(v+v_0) \to \mathcal{O}_S(C_{v'})(v+v_0) \to \mathcal{N}_{C_{v'}|S}(v+v_0) \to 0$$

and using the definitions of $a$ and $b$. Now since we already know

$${}_{v+v_0}H^{n-1}_{\mathfrak{m}}(K_{A_v}(n+4) \otimes_R I_v) \to L^{n-1}(C_v)_{v+v_0} \neq 0$$

is surjective, we get (ii) by the liaison-invariance of ${}_{v+v_0}H^{n-1}_{\mathfrak{m}}(K_{A_v}(n+4) \otimes_R I_v)$, and we conclude easily. □

COROLLARY 7.7. *Let $C \subset S \subset \mathbb{P}^{n+3}$ be two ACM subschemes, $S$ generically a complete intersection, and suppose $C$ is a Cartier divisor on $S$ of dimension $n = \dim C \geq 2$. Set $A = R/I$, $I = I_C$, let $i$ be an integer, $2 \leq i+1 \leq n$, and suppose $L^{i-1}(C)_{v_0} \neq 0$, or equivalently*

$$H^{i+1}(\mathcal{N}_S \otimes \mathcal{I}_{C|S}(v_0)) \neq 0$$

*for some $v_0$. Let $C_v \in |C+vH|$ be effective Cartier divisors and let $s-1$ be the largest integer $v$ such that*

$$H^{i+1}(\mathcal{N}_{C_v|S}(v+v_0)) \neq 0$$

*(such an $s \neq +\infty$ exists; $s = -\infty$ is, however, allowed). Then $C_v$, $v \geq s$, is not licci. Moreover, $C_v$ ($v \geq s$) and $C_{v'}$, for $v' \gg 0$, belong to different CI-liaison classes.*

PROOF. Consider the exact sequence

$$0 \to \mathcal{O}_S(v+v_0) \to \mathcal{O}_S(C_v)(v+v_0) \to \mathcal{N}_{C_v|S}(v+v_0) \to 0.$$

By taking cohomology, we get the existence of $s \neq +\infty$ because $\mathcal{O}_S(C_v) = \mathcal{O}_S(C)(v)$. Since

$$H^{i+1}(\mathcal{N}_S \otimes \mathcal{I}_{C|S}(v_0)) \cong H^{i+1}(\mathcal{N}_S \otimes \mathcal{I}_{C_v|S}(v+v_0)) \neq 0$$

we have by Proposition 7.3 a surjection

$${}_{v+v_0}H^{i-1}_{\mathfrak{m}}(K_{A_v}(n+4) \otimes_R I_v) \twoheadrightarrow L^{i-1}(C_v)_{v+v_0} \neq 0$$

for any $v \geq s$, where $I_v = I_{C_v}$ and $A_v = A/I_v$. Hence $C_v$ is not licci. Since, for $v' \gg 0$,

$${}_{v'+v_0}H^{i-1}_{\mathfrak{m}}(K_{A_v}(n+4) \otimes_R I_v) \cong L^{i-1}(C_v)_{v'+v_0} \cong L^{i-1}(C)_{v'-v+v_0} = 0$$

by Proposition 7.3 while ${}_{v'+v_0}H^{i-1}_{\mathfrak{m}}(K_{A_{v'}}(n+4) \otimes_R I_{v'}) \neq 0$, we are done. □

REMARK 7.8. Combining with Corollaries 7.6 and 7.7, we see the assumption $L^0(C)_{v_0} \neq 0$ for some $v_0$ implies that $C_v \in |C+vH|$ and $C_{v'}$ belong to different CI-liaison classes for large $v$ and $v'$ (provided $n \geq 1$; see Corollary 7.10 for the case $n = 0$). The mentioned assumption seems to be very weak, but there are a couple of cases where we know $L^0(C)$ always vanishes. In exactly the same cases we also know the conclusion is false, mainly by other arguments. First, if $S$ is a complete intersection, then the rank of $\bigoplus_i R(-q_i)$ in (7.1) is one, i.e. $\bigoplus_i \mathcal{I}_{C|S}(q_i+v) \cong \omega_S(n+4) \otimes \mathcal{I}_{C|S}(v)$ and we get $L^0(C) = 0$ by (7.2). This fits nicely with Remark 4.10 which states that in "standard" Basic Double Linkage, $C_v$ and $C_{v'}$ are CI-bilinked.

Secondly, if $C$ is a hypersurface section ($C \in |tH|$ for some $t$), one shows the surjectivity of the right map in (7.2) (i.e. $L^0(C) = 0$) by applying $\operatorname{Hom}_R(-, R)$ to the resolution (7.1). In this case, however, one may see directly that $C_v$ is licci for any $v$. Indeed, assume $C \in |tH|$. By Lemma 4.6, there exists some homogeneous polynomial $G$ of degree $t$ cutting out $C$. Let $F_1$ and $F_2$ be minimal generators of $I_S$. Then $(F_1, F_2)$ links $S$ to an ACM codimension two subscheme $S'$ with $\nu(I_{S'}) = \nu(I_S) - 1$ (here $\nu(-)$ denotes the number of minimal generators), and $(F_1, F_2, G)$ links $C$ to a subscheme $C'$ which is a hypersurface section of $S'$. Then in a finitely number of steps we link $C$ to a hypersurface section of a complete intersection, which is again a complete intersection. We remark that with the same proof we have something more general:

> If $S$ is licci (of any codimension) and $C$ is a global complete intersection in $R/I_S$ (i.e. $C$ has codimension $c$ in $S$ and is cut out on $S$ by a codimension $c$ complete intersection in $\mathbb{P}^N$) then $C$ is also licci.

Thirdly, if $C$ is arithmetically Gorenstein of the type described in Lemma 5.4 and Lemma 4.8 (iv), then the map

$$\bigoplus_i H^0(\mathcal{I}_{C|S}(q_i + v)) \to H^0(\omega_S(n+4) \otimes \mathcal{I}_{C|S}(v))$$

of (7.2), where $\mathcal{I}_{C|S} \cong \omega_S(h)$ for some $h$, must be surjective for any $v$. Indeed, the mentioned conclusion cannot hold because Lemma 4.8 (iv) implies that $C_v$ and $C_{v'}$ are arithmetically Gorenstein of codimension 3, and hence licci by [**73**].

REMARK 7.9. Let

$$e = 1 + \max\{\, v | H^{n+1}(\mathcal{I}_C(v)) \neq 0 \,\}.$$

Using the exact sequence

$$0 \to H^{n+2}_{\mathfrak{m}}(I) \to H^{n+2}_{\mathfrak{m}}(I_{C|S}) \to H^{n+3}_{\mathfrak{m}}(I_S) \to 0$$

we see that ${}_uH^{n+2}_{\mathfrak{m}}(I_{C|S}) = 0$ for any $u \geq \max\{a, e\}$ because

$$a = 1 + \max\{v | H^{n+2}(\mathcal{I}_S(v)) \neq 0\}.$$

So the number $t$ of Corollary 7.6 (i.e. a bound for which the corollary is true) can be taken such that $t + s(C|S) \geq \max\{a, e\}$, i.e. as

$$t = \max\{a, e\} - s(C|S), \tag{7.10}$$

where $s(C|S)$ is the minimal degree of the generators of $I_{C|S}$. Note that the minimal degree $s(C)$ of the generators of $I$ satisfy $s(C) \leq s(C|S)$, i.e. the corollary holds putting $t = \max\{a, e\} - s(C)$ as well.

Since the considerations in Remark 7.9 hold for $n = 0$, we will prepare a corollary using Proposition 7.2 (ii). Now supposing $n = 0$, we *claim* that

$$t = \max\{a, e\} - s(C|S), \ u \geq t \text{ and } v \geq a - \max\{a, e\}$$

implies ${}_u\operatorname{Ext}^i_R(I_{C_v|S}, H^2_{\mathfrak{m}}(I_v)) = 0$ for $i = 0, 1$. Indeed since $s(C_v|S) = s(C|S) + v$ and $u + s(C|S) \geq \max\{a, e\}$, we get $u + s(C_v|S) \geq a$ and hence

$${}_u\operatorname{Hom}_R(I_{C_v|S}, H^3_{\mathfrak{m}}(I_S)) = 0.$$

It suffices therefore to prove ${}_u\operatorname{Ext}^i_R(I_{C_v|S}, H^2_{\mathfrak{m}}(I_{C_v|S})) = 0$ for $i = 1, 2$. This, however, follows from $I_{C_v|S} \cong I_{C|S}(-v)$, $H^2_{\mathfrak{m}}(I_{C_v|S}) \cong H^2_{\mathfrak{m}}(I_{C|S}(-v))$ and from ${}_uH^2_{\mathfrak{m}}(I_{C|S}) = 0$ for $u \geq \max\{a, e\}$ because the minimal relation degree $r(C|S)$ in a minimal $R$-resolution of $I_{C|S}$ satisfies $r(C|S) > s(C|S)$. Now, combining the proven claim with Proposition 7.2 (ii), we have

COROLLARY 7.10. *Let $C \subset S$ be two ACM subschemes in $\mathbb{P}^3$, $S$ generically a complete intersection, and suppose $C$ is a Cartier divisor on $S$ of dimension $n = \dim C = 0$. Let $t = \max\{a, e\} - s(C|S)$ and $C_v \in |C + vH|$ be effective Cartier divisors, and suppose $L^0(C)_{v_0} \neq 0$ for some $v_0$. Then $C_v$ and $C_{v'}$ are not licci and they belong to different CI-liaison classes for any $v > v' \geq \min\{0, a - e\}$ such that $v' \geq t - v_0$.*

EXAMPLE 7.11. The Castelnuovo surface $S \subset \mathbb{P}^4$ can be realized as the blow-up of $\mathbb{P}^2$ in 8 general points, leading to $\mathrm{Pic}(S) \cong \mathbb{Z}^{\oplus 9}$, which allows us to write $H = (4; 2, 1^7)$ and $-K = (3; 1^8)$; see Chapter 8 of this paper. Moreover one knows

$$0 \to R(-4)^{\oplus 2} \to R(-3)^{\oplus 2} \oplus R(-2) \to I_S \to 0.$$

a) Consider a curve $C$ on $S$ of the system $(1; 1, 0^7)$ of degree $d = C \cdot H = 2$ and genus $g = 0$. If $C_v$ belongs to $(4v + 1; 2v + 1, v^7)$, i.e. to $|C + vH|$, *we claim that $C_v$ and $C_{v'}$ are not licci and belong to different CI-liaison classes provided $v > v' \geq 2$.* Indeed, using Corollary 7.6 and (7.10), this is just an exercise in computing the constants $a$, $b$ etc. Since $C$ and $S$ are rational, we get $a = 0$ and $e = 0$, and we can take $t = -1$ by (7.10). Moreover, since $-C + K = -H$ modulo linear equivalence, we get

$$H^1(\mathcal{O}_S(C)(v)) \cong H^1(\mathcal{O}_S(-C + K)(-v))^\vee = H^1(\mathcal{O}_S(-v - 1))^\vee = 0$$

for any $v$ because $S$ is an ACM surface. Hence $b = -\infty$ by definition. To conclude by Corollary 7.6 it remains to compute $v_0$, i.e. to consider $v$'s where the morphism

$$\bigoplus_i H^0(\mathcal{I}_{C|S}(q_i + v)) = H^0(\mathcal{I}_{C|S}(4 + v))^{\oplus 2} \to H^0(\omega_S(5) \otimes \mathcal{I}_{C|S}(v)) = H^0(\mathcal{O}_S(4 + v))$$

is *not* surjective. It is obviously not surjective for $v = -4$, nor for $v = -3$ because we easily show $h^0(\mathcal{I}_{C|S}(1)) = 2$ and $h^0(\mathcal{O}_S(1)) = 5$. Hence we can use Corollary 7.6 with $v_0 = -3$, and we get the claim.

b) Let $C$ be a curve of the system $(1; 0^8)$ of degree $d = C \cdot H = 4$ and genus $g = 0$. If $C_v$ belongs to $(4v + 1; 2v, v^7)$, we similarly prove that *$C_v$ and $C_{v'}$ are not licci and belong to different CI-liaison classes provided $v > v' \geq 1$, $v \geq 3$.* Indeed $a = 0$, $e = 0$ as above, i.e. $t = -2$ by (7.10) while $b \leq 0$ because, on an ACM surface, $C \cdot K < 0$ implies $H^1(\mathcal{O}_S(C)(v)) \hookrightarrow H^1(\mathcal{N}_{C|S}(v)) \cong H^0(\mathcal{O}_S(K) \otimes \mathcal{O}_C(-v))^\vee = 0$ for all $v \geq 0$. Finally $H^0(\mathcal{I}_{C|S}(4 + v))^{\oplus 2} \to H^0(\omega_S(5) \otimes \mathcal{I}_{C|S}(v))$ is obviously not surjective for $v = -3$ ($H^0(\mathcal{I}_{C|S}(1)) = 0$), i.e. we can take $v_0$ in Corollary 7.6 to be $-3$ and we are done.

Note that the minimal resolution of $I_C$ coincides with the minimal resolution of the rational quartic curve considered in Example 6.14, i.e. $C = C_0$ is not licci, and neither is $C_1$ because one shows that the minimal resolution of $I_{C_1}$ is almost linear in the sense of Example 6.14 with $(a, b, c_1, c_0, s) = (5, 12, 7, 1, 2)$ and $(d, g) = (9, 5)$. Hence $l(C_1)_{-2} - l(C_1)_{-3} = d + g - 1 - ac_0 = 8$, while $l(C_0)_{-2} - l(C_0)_{-3} = 3$ by the same formula. Continuing the computations of Example 6.14, we immediately get $l(C_0)_v = 0$ for $v \geq -1$. Combining with Corollary 7.6 (i) which implies $l(C_v)_{v-3} \neq 0$ for $v \geq 1$, we see that *$C_v$ and $C_{v'}$ are not licci and belong to different CI-liaison classes provided $v > v' \geq 0$* (also for $C_1$ and $C_2$ belonging to different CI-liaison classes because one may prove $l(C_2)_{-2} - l(C_2)_{-3} = 6$).

c) Suppose $C$ is a section of $(0; 0^7, -1)$, and so has degree $d = 1$. Then $a = 0$ ($S$ is rational), $e \leq 0$ ($C$ is rational), $b \leq 0$ ($C \cdot K < 0$), $t = -1$ by (7.10), $v_0 = -4$ ($h^0(\omega_S(5) \otimes \mathcal{I}_{C|S}(-4)) \neq 0$), i.e. *$C_v$ and $C_{v'}$ are not licci and belong to different CI-liaison classes provided $v > v' \geq 3$.*

d) Suppose $C$ is a section of $(4;3,1^7)$, and so has degree $d=3$ and genus $g=0$. Then $a=0$ ($S$ is rational), $e \leq 0$ ($C$ is rational), $b \leq 0$ ($C \cdot K < 0$), $t=-1$ by (7.10), $v_0 = -3$ ($h^0(\mathcal{I}_{C|S}(1)) = 1$ and $h^0(\omega_S(5) \otimes \mathcal{I}_{C|S}(-3)) = 3$), i.e. *$C_v$ and $C_{v'}$ are not licci and belong to different CI-liaison classes provided $v > v' \geq 2$.*

EXAMPLE 7.12. Let $S$ be a smooth ACM curve of degree $d=5$ and genus $g=2$ in $\mathbb{P}^3$, in which case one knows the homogeneous ideal $I_S$ in $R = k[x_0,x_1,x_2,x_3]$ satisfies

$$0 \to R(-4)^{\oplus 2} \to R(-3)^{\oplus 2} \oplus R(-2) \to I_S \to 0,$$

and let $C$ be a set of $n$ reduced points on $S$ with Hilbert function $H = (h_0, h_1, h_2, ...)$, $h_i = \dim A_i$. To compute $L^0(C)_v$, we consider the right map in the complex (7.2). If $n \leq 5$, we have by the Riemann-Roch Theorem

$$\begin{aligned} h^0(\omega_S(4) \otimes \mathcal{I}_{C|S}(-3)) &\geq \chi(\omega_S(1) \otimes \mathcal{O}_S(-C)) \\ &= 2g-2+d-n+1-g \\ &= 6-n \\ &> 0. \end{aligned}$$

Therefore if $h_1 = 4$ (hence $n \geq 4$ and $h^0(\mathcal{I}_C(1)) = 0$), or if $h_1 = 3$ and $n = 3$ (hence $H = (1,3,3,3..)$ and $h^0(\mathcal{I}_C(1)) = 1$), then the right-hand map in (7.2) is not surjective for $v = 3$, i.e. $L^0(C)_{-3} \neq 0$. It follows from Corollary 7.10 that *$C_v$ and $C_{v'}$ are not licci and that they belong to different CI-liaison classes for $v > v' \gg 0$.* Noting that the left-hand term of the minimal resolution of $I_S$ leads to $a=1$, we have in particular

- If $h_1 = 3$ and $n = 3$, then $e = 1$ and $s(C|S) = 1$, i.e. $C_v$ and $C_{v'}$ are not licci and belong to different CI-liaison classes provided $v > v' \geq 3$.
- If $h_1 = 4$ and $n = 4$, then $e = 1$ and $s(C|S) = 2$, i.e. $C_v$ and $C_{v'}$ are not licci and belong to different CI-liaison classes provided $v > v' \geq 2$.
- If $H = (1,4,5,5,..)$, then $e = 2$ and $s(C|S) = 2$, i.e. $C_v$ and $C_{v'}$ are not licci and belong to different CI-liaison classes provided $v > v' \geq 3$.

REMARK 7.13. In contrast to the situation with G-liaison, it is interesting to note that linearly equivalent divisors on an arithmetically Cohen-Macaulay scheme need not be in the same CI-liaison class. The scheme arising in the second part of Example 5.8 is linearly equivalent on a Castelnuovo surface to a Gorenstein scheme. It is easy to check that they have different CI-liaison invariants. The very simplest example is of points lying on the ACM curve in $\mathbb{P}^3$ given in Example 7.12. More precisely, let $Z \subset \mathbb{P}^3$ consist of the union of three points on a line and two points on a different line. The minimal free resolution of the homogeneous ideal of $Z$ is

$$0 \to R(-4)^{\oplus 2} \oplus R(-5) \to R(-3)^{\oplus 6} \oplus R(-4)^{\oplus 2} \to R(-2)^{\oplus 5} \oplus R(-3) \to I_Z \to 0.$$

Let $L \subset \mathbb{P}^3$ be a general line and let $Y = L \cup Z$. Let $F$ and $G$ be two general forms from the homogeneous ideal of $Y$ of degrees two and three respectively. Link $L$ to a curve $C$ using the complete intersection $(F,G)$. The curve $C$ will have degree five and genus two and it contains $Z$. The general element, $Z'$, in the linear system $|Z|$ on $C$ is arithmetically Gorenstein with minimal free resolution

$$0 \to R(-5) \to R(-3)^{\oplus 5} \to R(-2)^{\oplus 5} \to I_{Z'} \to 0.$$

The scheme $Z'$ is CI-linked to a complete intersection but $Z$ itself is not CI-linked to a complete intersection since $H^0_{\mathfrak{m}}(K_{R/I_Z} \otimes_R I_Z) \neq 0$.

In Examples 7.5, 7.11 a) and the first case of 7.12, $C$ was a section of a twist of the canonical sheaf $\omega_S$. Indeed, in any non-trivial case where $\mathcal{O}_S(C) \cong \omega_S(h)$ for some $h$, we can prove that $L^0(C)_{h-n-4} \neq 0$; hence we have the following general result.

PROPOSITION 7.14. *Let $C \subset S$ be two ACM schemes, $n = \dim C \geq 0$, where $S$ is generically, but not globally, a complete intersection in $\mathbb{P}^{n+3}$. Moreover, suppose $C$ is a Cartier divisor on $S$ satisfying $\mathcal{O}_S(C) \cong \omega_S(h)$ for some $h$, and let $C_v \in |C + vH|$ be effective Cartier divisors. Then $C_v$ and $C_{v'}$ are not licci and they belong to different CI-liaison classes for any $v > v' \gg 0$.*

PROOF. Since $a$ is the largest number satisfying $H^{n+1}(\mathcal{O}_S(a-1)) \neq 0$, we get $a - 1 \geq q_i - n - 4$ by (7.1). Moreover, by duality and the definition of $a$, we have $H^0(\omega_S(-a+1)) \neq 0$, i.e. we have injections

$$\mathcal{O}_S \hookrightarrow \omega_S(-a+1) \hookrightarrow \omega_S(-q_i + n + 4)$$

for any $i$, whose composition is not an isomorphism because $S$ is not a complete intersection. Now, $\omega_S(h) \cong \mathcal{O}_S(C)$ implies

$$\mathcal{O}_S \hookrightarrow \mathcal{O}_S(C)(-h - q_i + n + 4), \quad \text{i.e.} \quad \mathcal{I}_{C|S}(q_i + h - n - 4) \underset{\neq}{\hookrightarrow} \mathcal{O}_S$$

which shows that $H^0(\mathcal{I}_{C|S}(q_i + h - n - 4)) = 0$ for any $i$. Using the complex (7.2) with $v = h - n - 4$, we have

$$\bigoplus_i H^0(\mathcal{I}_{C|S}(q_i + v)) \to H^0(\omega_S(n+4) \otimes \mathcal{I}_{C|S}(h-n-4)) \cong H^0(\omega_S(h) \otimes \mathcal{O}_S(-C)) \cong k$$

i.e. $L^0(C)_{h-n-4} \neq 0$, and we conclude by Corollary 7.6,Corollary 7.7 and Corollary 7.10. ☐

It is possible to modify the approach above to treat ACM subschemes $C \subset S \subset \mathbb{P}^N$, $N = n+c$, of any codimension $c \geq 3$ and $c-1$, respectively. Reconsidering the proof of Proposition 7.2 we see that $\ker \delta_C$ and $\operatorname{coker} \delta_C$, which are liaison-invariant by Theorem 6.1 (and under a local complete intersection assumption isomorphic to ${}_{-v}H^i_{\mathfrak{m}}(K_A \otimes_R I)^\vee$ for $i = 1$ and $0$ respectively), fit into

$$(7.11) \qquad \begin{gathered} H^n(\mathcal{N}_{C|S}(v)) \to \ker \delta_C \to \ker \phi_{C|S} \to 0 \quad (n > 0), \\ (\text{resp. } \operatorname{coker} \delta_C \xrightarrow{\sim} \operatorname{coker} \phi_{C|S}, \quad n \geq 0) \end{gathered}$$

provided ${}_v\operatorname{Hom}_R(I, H^{n+2}_{\mathfrak{m}}(I)) \to {}_v\operatorname{Hom}_R(I_S, H^{n+2}_{\mathfrak{m}}(I))$ is injective (resp. an isomorphism). If we let

$$\rho'_{C|S} : H^{n+1}(\mathcal{N}_S \otimes \mathcal{I}_{C|S}(v)) \to {}_v\operatorname{Ext}^{n+2}_{\mathfrak{m}}(I_S, I_{C|S}) \cong {}_v\operatorname{Hom}_R(I_S, H^{n+2}_{\mathfrak{m}}(I_{C|S}))$$

and if we suppose

$$(7.12) \qquad H^n(\mathcal{N}_S(v)) = 0 \text{ and } H^{n+1}(\mathcal{N}_S(v)) \xrightarrow{\delta_S} {}_v\operatorname{Hom}_R(I_S, H^{n+3}_{\mathfrak{m}}(I_S)) \text{ injective}$$

(resp. replace (7.12) by: $\delta_S$ an isomorphism), then the diagram in the proof of Proposition 7.2 which contains $H^i(\mathcal{N}_S(v))$ for $i = n$ and $n+1$, shows

$$\ker \phi_{C|S} \cong \ker \rho'_{C|S} \quad (\text{resp. } \operatorname{coker} \phi_{C|S} \cong \operatorname{coker} \rho'_{C|S}).$$

Combining with (7.11) we get the variation of Proposition 7.2 we have in mind. Note that defining

$$Li^{n-1}(C)_v = \ker \rho'_{C|S} \quad \text{and} \quad Li^n(C)_v = \operatorname{coker} \rho'_{C|S},$$

then these groups obviously satisfy $Li^j(C)_v \cong Li^j(C_t)_{v+t}$, for any $C_t \in |C + tH|$, and they coincide with $L^j(C)$ in codimension 3 (cf. the proof of Proposition 7.2). Note also that the assumptions of (7.12) are satisfied and that $\delta_S$ is an isomorphism if $S \subset \mathbb{P}^{n+c}$ is licci by Corollary 6.5. Putting all this together we get the following result (with a proof so close to the proof of Corollary 7.6 that we skip the details).

PROPOSITION 7.15. *Let $C \subset S$ be two ACM subschemes of $\mathbb{P}^{n+c}$, $S = \operatorname{Proj}(B)$ generically a complete intersection subscheme in $\mathbb{P}^{n+c}$ satisfying $H^2(R, B, B)^{\sim} = 0$ and $C$ a Cartier divisor on $S$ of dimension $n = \dim C \geq 1$ (resp. $n \geq 0$), and let $l$ (resp. $l'$) be an integer such that $H^n(\mathcal{N}_S(v)) = 0$ and $\delta_S$ is injective (resp. $\delta_S$ is an isomorphism) for any $v \geq l$ (resp. $v \geq l'$). Let $a$, $b$ and $e$ be as in Corollary 7.6 and Remark 7.9, let $t = \max\{a, e\} - s(C|S)$, and suppose $Li^{n-1}(C)_{v_0} \neq 0$ (resp. $Li^n(C)_{v_0} \neq 0$) for some $v_0$. If $C_v \in |C + vH|$ are effective Cartier divisors, then*

(i) *$C_v$ is not licci provided $v \geq \max\{t, l\} - v_0$ (resp. $v \geq \max\{t, l'\} - v_0$ and $v \geq \min\{0, a - e\}$).*
*Indeed,*

$$ {}_{v+v_0}H^{n-1}_{\mathfrak{m}}(K_{A_v}(n+c+1) \otimes_R I_v) \neq 0 $$

$$ (\textit{resp. } {}_{v+v_0}H^{n}_{\mathfrak{m}}(K_{A_v}(n+c+1) \otimes_R I_v) \cong Li^n(C)_{v_0}). $$

(ii) *$C_v$ and $C_{v'}$ belong to different CI-liaison classes for any $v$ and $v'$ satisfying*

$$ v > v' \geq \max\{t, l\} - v_0, \quad v \geq a - v_0 \quad \textit{and} \quad v + v' \geq b - v_0 $$

$$ (\textit{resp. } v > v' \geq \max\{t, l'\} - v_0, \;\; v' \geq \min\{0, a - e\}). $$

*In fact* ${}_{u+v_0}H^{n-1}_{\mathfrak{m}}(K_{A_u}(n+c+1) \otimes_R I_u) \cong Li^{n-1}(C)_{v_0}$ *for* $u \geq v - 1$.

CHAPTER 8

# Glicci curves on Arithmetically Cohen-Macaulay surfaces

The goal of this chapter is to see, again, that often Gorenstein liaison behaves better than complete intersection liaison. We will see that there is only one G-liaison class containing ACM curves $C \subset \mathbb{P}^4$ lying on a "general" smooth, rational, ACM surface $X$ of $\mathbb{P}^4$; even more, we will see that all ACM curves $C \subset \mathbb{P}^4$ lying on a "general" smooth, ACM, rational surface $X \subset \mathbb{P}^4$ are glicci (i.e. in the Gorenstein liaison class of a complete intersection). We want to stress that this result drastically differs from the ones obtained in Chapter 6 and Chapter 7. Indeed, it follows from Corollary 7.6 that on any non complete intersection, ACM, smooth, rational surface $S \subset \mathbb{P}^4$ there are infinitely many ACM curves all of them belonging to different CI-liaison classes.

The fact that all "standard" determinantal ideals are glicci and that all ACM curves $C \subset \mathbb{P}^4$ lying on a "general" smooth ACM rational surface $X \subset \mathbb{P}^4$ are glicci, suggests to us the following question, which would generalize Gaeta's codimension two result even further than what we already saw in Chapter 3 (see also Remark 3.7):

QUESTION 8.1. *In any codimension, is there only one Gorenstein liaison class containing ACM schemes?*

The first goal of this chapter is to see that on ACM surfaces there are only finitely many G-liaison classes with fixed Rao module.

DEFINITION 8.2. Let $X$ be a smooth subscheme of $\mathbb{P}^n$. We will say that an effective divisor $C$ on $X$ is minimal if there is no effective divisor in the linear system $|C - H|$, where $H$ is the hyperplane section divisor.

We will first consider smooth rational surfaces. Recall that if $X$ is a smooth rational surface, then either $X$ is obtained by blowing up $\mathbb{P}^2$ at $s \geq 0$ different points, or by blowing up a Hirzebruch surface $X_e$, $e \geq 0$, at $s \geq 0$ different points.

Assume that $X$ is obtained by blowing up $\mathbb{P}^2$ at $s \geq 0$ different points. Taking the linear equivalence classes of the inverse image $E_0$ of a line in $\mathbb{P}^2$ and $E_i$ ( the exceptional divisor), $i = 1, ..., s$, as a basis for $\mathrm{Pic}(X)$, we can associate a curve with an $(s+1)$-tuple $(a; b_1, ..., b_s)$. With this notation, the canonical divisor is $K = (-3; -1, ..., -1)$, $E_0^2 = 1 = -E_1^2 = \cdots = -E_s^2$, and $E_iE_j = 0$ if $i \neq j$.

For the Hirzebruch surface $X_e \cong \mathbb{P}(\mathcal{O}_{\mathbb{P}^1} \oplus \mathcal{O}_{\mathbb{P}^1}(-e))$, $e \geq 0$, we denote the standard basis of $\mathrm{Pic}(X_e) \cong \mathbb{Z} \oplus \mathbb{Z}$ by $C_0$ and $f$: they correspond to sections and $p$-fibers respectively of the natural projection $p : X_e \to \mathbb{P}^1$; and they verify $C_0^2 = -e$, $f^2 = 0$ and $C_0 f = 1$: ([**31**] V, Proposition 2.3). Furthermore, the canonical divisor is $K = -2C_0 - (e+2)f$, $K^2 = 8$ and $-K$ is effective.

Assume that $X$ is obtained blowing up a Hirzebruch surface $X_e$ at $s \geq 0$ different points and let $\pi : X \longrightarrow X_e$ be the blow up. The Picard group of $X$ is generated by $\pi^*C_0$, $\pi^*f$, $E_1$,...,$E_s$, with $(\pi^*C_0)^2 = -e$, $\pi^*C_0\pi^*f = 1$, $(\pi^*f)^2 = 0$, $\pi^*C_0E_i = \pi^*fE_i = 0$, for $1 \leq i \leq s$, $1 = -E_1^2 = \cdots = -E_s^2$, and $E_iE_j = 0$ if $i \neq j$. The canonical divisor has the form

$$K_X = \pi^*K_{X_e} + \sum_{i=1}^{s} E_i = -(e+2)\pi^*f - 2\pi^*C_0 + \sum_{i=1}^{s} E_i$$

([**31**] V, Propositions 3.2 and 3.3). For simplicity we will write $C_0$ and $f$ instead of $\pi^*C_0$ and $\pi^*f$ and we will identify a curve with an $(s+2)$-uple $(a, b; b_1, ..., b_s)$.

Let us see that there are only finitely many G-liaison classes containing ACM curves $C \subset \mathbb{P}^n$ lying on a smooth ACM rational surface. We first re-state a special case of Corollary 5.14:

LEMMA 8.3. *Let $X$ be a generic complete intersection, ACM subscheme of $\mathbb{P}^n$. Let $C$ be an effective divisor on $X$ and let $H$ be the hyperplane section divisor. Then $C$ and $C+H$ are in the same Gorenstein liaison class.*

PROPOSITION 8.4. *Let $X \subset \mathbb{P}^n$ be a smooth, ACM rational surface. There exist only finitely many G-liaison classes of ACM curves $C \subset X$.*

PROOF. By Lemma 8.3 we only need to check that on $X$ there are only finitely many minimal ACM curves. If $X = \mathbb{P}^2$, the result is obvious. Otherwise, we distinguish two cases:

Case A: $X$ is obtained by blowing up $\mathbb{P}^2$ at $s \geq 1$ different points and we assume that $X$ is embedded in $\mathbb{P}^n$ by means of the divisor $H = aE_0 - \sum_{i=1}^{s} b_iE_i$. Let $C$ be an (effective) divisor on $X$ such that, viewed as a subscheme of $\mathbb{P}^n$, $C$ is ACM. Without loss of generality we can associate to $C$ the divisor $(a\nu + \alpha)E_0 - \sum_{i=1}^{s}(b_i\nu + \beta_i)E_i = \nu H + \alpha E_0 - \sum_{i=1}^{s} \beta_i E_i$ with $0 \leq \alpha \leq a-1$ and $\beta_i \in \mathbb{Z}$. Now, the result easily follows from the next claim which gives us lower and upper bounds for $\beta_i$, $i = 1, ..., s$.

*Claim 1:* If $\alpha = 0$ then $-1 \leq \beta_i \leq 1$ and at most one $\beta_i$ is equal to $-1$. If $1 \leq \alpha \leq a-1$ then $0 \leq \beta_i \leq \max\{1, \alpha - 1\}$. Furthermore, if $\alpha \geq 3$ at most one $\beta_i$ is equal to $\alpha - 1$.

*Proof of Claim 1:* We will first discuss the upper bounds. Assume that there exists $\beta_i > \max\{\alpha - 1, 1\}$ and consider the exact sequence:

$$0 \longrightarrow \mathcal{O}_X(-E_i) \longrightarrow \mathcal{O}_X \longrightarrow \mathcal{O}_{E_i} \longrightarrow 0. \tag{8.1}$$

Since $C$ is an ACM subscheme of $\mathbb{P}^n$, we have

$$0 = H^1(\mathbb{P}^n, \mathcal{I}_C(\nu)) = H^1(\mathcal{O}_X(-C + \nu H)).$$

By Serre duality, we have $H^2(\mathcal{O}_X(-C+\nu H - E_i)) = H^0(\mathcal{O}_X(C - \nu H + E_i + K)) = H^0(\mathcal{O}_X((\alpha-3)E_0 - (\beta_i - 2)E_i - \sum_{j\neq i}(\beta_j - 1)E_j)) = 0$. Twisting the exact sequence (8.1) and taking cohomology we get:

$$0 = H^1(\mathcal{O}_X(-C+\nu H)) \longrightarrow H^1(\mathcal{O}_{E_i}(-C+\nu H)) \longrightarrow H^2(\mathcal{O}_X(-C+\nu H - E_i)) = 0.$$

Therefore, $H^1(\mathcal{O}_{E_i}(-C+\nu H)) = 0$ which is a contradiction because $E_i$ is a rational curve of degree $b_i$ and $E_i(-C + \nu H) = -\beta_i < -\max\{\alpha - 1, 1\} \leq -1$. A similar argument proves that if $\alpha \geq 3$ then at most one $\beta_i = \alpha - 1$.

Concerning the lower bounds, if $1 \leq \alpha \leq a-1$ we have

$$0 = H^1(\mathcal{O}_X(-C + \nu H)) = H^0(\mathcal{O}_X(-C + \nu H + E_i)).$$

From the exact sequence (8.1) we get $H^0(\mathcal{O}_{E_i}(-C+\nu H+E_i))=0$ and hence $\beta_i \geq 0$. If $\alpha=0$, similar argument gives us that $\beta_i \geq -1$ and at most one $\beta_i$ is equal to $-1$.

Case B: $X$ is obtained by blowing up a Hirzebruch surface $X_e$, $e \geq 0$, at $s \geq 0$ different points and we assume that $X$ is embedded in $\mathbb{P}^n$ by means of the divisor $H=af+bC_0-\sum_{i=1}^s b_iE_i$. Let $C$ be an (effective) divisor on $X$ such that, viewed as a subscheme of $\mathbb{P}^n$, $C$ is ACM. Without loss of generality we can associate to $C$ the divisor $(a\nu+\alpha)f+(b\nu+\beta)C_0-\sum_{i=1}^s(b_i\nu+\beta_i)E_i=\nu H+\alpha f+\beta C_0-\sum_{i=1}^s\beta_iE_i$ with $0\leq \alpha \leq a-1$ and $\beta,\beta_i \in \mathbb{Z}$. Now, the result easily follows from the next claim which gives us lower and upper bounds for $\beta$ and $\beta_i$, $i=1,...,s$.

*Claim 2:* If $\alpha \geq 2$ then $\beta_i, \beta-1 \geq 0$; if $0 \leq \alpha \leq 1$ then $\beta_i,\beta \geq \alpha-1$. If $0 \leq \alpha \leq e$ then $\beta,\beta_i \leq 1$; if $\alpha=e+1$ then $\beta_i \leq 1$ and $\beta \leq 2$; and if $\alpha \geq e+2$ then $\beta_i \leq 4\alpha$ and $\beta \leq 2\alpha$.

*Proof of the Claim 2:* It follows from the Riemann-Roch Theorem and the exact cohomology sequences associated to the exact sequences:

$$0 \longrightarrow \mathcal{O}_X(-E_i) \longrightarrow \mathcal{O}_X \longrightarrow \mathcal{O}_{E_i} \longrightarrow 0$$

$$0 \longrightarrow \mathcal{O}_X(-f) \longrightarrow \mathcal{O}_X \longrightarrow \mathcal{O}_f \longrightarrow 0.$$

□

Now we will generalize Proposition 8.4 to arbitrary ACM surfaces and arbitrary G-liaison classes. To this end, the following lemma will be very useful.

LEMMA 8.5. *Let $X \subset \mathbb{P}^n$ be a generic complete intersection surface. If $H^1(X,\mathcal{O}_X)=0$ then* Pic$(X)$ *is finitely generated.*

PROOF. We consider the exponential map $e:\mathcal{O}_X \longrightarrow \mathcal{O}_X^*$ defined by $e(f):=exp(2\pi i f)$ and the corresponding exact sequence:

$$0 \longrightarrow \mathbb{Z} \longrightarrow \mathcal{O}_X \longrightarrow \mathcal{O}_X^* \longrightarrow 0.$$

From the vanishing $H^1(X,\mathcal{O}_X)=0$, we deduce that Pic$(X)$ is a subgroup of $H^2(X,\mathbb{Z})$ and therefore Pic$(X)$ is finitely generated. □

PROPOSITION 8.6. *Let $X$ be a smooth, ACM (hence integral) surface in $\mathbb{P}^n$ and let $M$ be a graded $R$-module of finite length. Then there exist only finitely many G-liaison classes of curves $C \subset X$ with Rao module $M(C) \cong M$ (up to twist).*

PROOF. Note that the conditions that $X$ be smooth and ACM together imply that $X$ is integral. Since $H^1(X,\mathcal{O}_X)=0$, the Picard group Pic$(X)$ is finitely generated. We choose a system of generators $\{L_1,...,L_r\}$ of Pic$(X)$ and we assume that $X$ is embedded in $\mathbb{P}^n$ by the divisor $H=\sum_{i=1}^r h_iL_i$. Set $K=\sum_{i=1}^r k_iL_i$. Let $\Delta=(\delta_{ij})$ be the symmetric matrix associated to the bilinear pairing Pic$(X)\times$Pic$(X) \longrightarrow \mathbb{Z}$.

Again, by Lemma 8.3 we only need to check that on $X$ there are only finitely many minimal curves $C$ (up to linear equivalence) with Rao module $M$ (up to shift). Let $C$ be an effective divisor on $X$ with $M(C) \cong M$ (up to shift), and assume that $C-H$ is not effective. Set $n_t:=\dim_k M(C)_t$. We associate to $C$ the divisor $\nu H+\sum_{i=1}^r c_iL_i$. The proof follows from the following claim:

*Claim:* there exist upper and lower bounds for $c_i$, $i=1,...,r$.

*Proof of the Claim:* From now on, we will assume that the divisors $L_i$, $i = 1, ..., r$, are effective; otherwise, we replace $L_i$ by $L_i + \alpha_i H$ for a suitable choice of $\alpha_i$. We set:

$$d_i := \deg(L_i) = \sum_{j=1}^{r} h_j \delta_{ij}$$

and

$$g_i := p_a(L_i) = 1 + \frac{L_i(L_i + K)}{2} = 1 + \frac{\delta_{ii} + \sum_{j=1}^{r} k_j \delta_{ji}}{2}.$$

We consider the exact sequences:

$$(8.2) \qquad 0 \longrightarrow \mathcal{O}_X(-L_i) \longrightarrow \mathcal{O}_X \longrightarrow \mathcal{O}_{L_i} \longrightarrow 0 \qquad i = 1, ..., r.$$

We distinguish two cases:

(i) There exists $i$ such that $-C + L_i$ is effective;
(ii) For all $i$, $-C + L_i$ is not effective.

For case (i), we have $(-C + L_i)H \geq 0$, or equivalently, $C \cdot H \leq L_i \cdot H$ (recall that the intersection number $C \cdot H$ is just equal to the degree of $C$ in $\mathbb{P}^n$). Since, for any integer $d$, the set of effective divisors $D$ on $X$ with $D \cdot H = d$ is a finite set (cf. **[31]** Chapter V, Exercise 1.11), we conclude that there are only finitely many minimal curves $C$ on $X$ with Rao module $M$ (up to shift) and $-C + L_i$ effective.

For case (ii), from the exact cohomology sequence associated to (8.2), we deduce

$$h^0(\mathcal{O}_{L_i}(-C + L_i)) \leq h^1(\mathcal{O}_X(-C)) = \dim_k M(C)_0 = n_0.$$

Therefore,

$$\begin{aligned} (-C + L_i)L_i + 1 - g_i &= \chi(\mathcal{O}_{L_i}(-C + L_i)) \\ &\leq h^0(\mathcal{O}_{L_i}(-C + L_i)) \\ &\leq n_0 \end{aligned}$$

or equivalently

$$-\sum_{j=1}^{r} \delta_{ij} c_j + \delta_{ii} + 1 - g_i \leq n_0.$$

This gives the desired lower bound.

On the other hand, for any integer $i$, $1 \leq i \leq r$, we define

$$q_i = \min\{t \in \mathbb{Z} | - L_i - K + (t-1)H \text{ is effective}\}$$

and

$$q = \max_{1 \leq i \leq r} \{q_i\}.$$

(Notice that the definition of $q$ does not depend on $C$.) We note that $C + L_i - qH + K$ is not effective. Indeed, if $C + L_i - qH + K$ were effective, since $-L_i - K + (q-1)H$ is effective (by definition of $q$), we deduce that

$$C + L_i - qH + K - L_i - K + (q-1)H = C - H$$

is effective, which contradicts the minimality of $C$. Therefore,

$$0 = H^0(\mathcal{O}_X(C + L_i - qH + K)) = H^2(\mathcal{O}_X(-C - L_i + qH)).$$

The exact cohomology sequence associated to (8.2) together with the fact that $C + L_i - qH + K$ is not effective gives us

$$h^1(\mathcal{O}_{L_i}(-C + qH)) \leq h^1(\mathcal{O}_X(-C + qH)) = n_q$$

or, equivalently,

$$\sum_{j=1}^{r} \delta_{ij} c_j - qd_i - 1 + g_i \leq n_q,$$

which gives the desired upper bound. □

REMARK 8.7. Proposition 8.6 would not be true if we only assume that $X$ is ACM and satisfies $G_1$, as we have done for most of this paper. Indeed, consider the surface $X$ of Example 5.18. Example 5.2 of [**30**] shows that $\mathrm{Pic}(X) = \mathbb{Z}$, generated by the hyperplane section, so $\mathrm{Pic}(X)$ does not contain all the effective curves on $X$. Instead, we would have to use $\mathrm{APic}(X)$, which Example 5.6 of [**30**] describes in a simple way (and which is not finitely generated, as is clear from our Example 5.18).

Our next goal is to describe the minimal effective divisors $C$ on a smooth ACM non-degenerate rational surface $X \subset \mathbb{P}^4$ which as subschemes of $\mathbb{P}^4$ are ACM and we will check that all of them are glicci. According to the classification (up to isomorphism) of smooth, ACM, rational surfaces $X \subset \mathbb{P}^4$, we distinguish 4 cases ($char(k) = 0$):

- Cubic scroll: $X = Bl_{\{p_1\}}(\mathbb{P}^2)$ embedded in $\mathbb{P}^4$ by means of the linear system $|2E_0 - E_1|$ and $\deg(X) = 3$.
- Del Pezzo surface: $X = Bl_{\{p_1,\ldots,p_5\}}(\mathbb{P}^2)$ embedded in $\mathbb{P}^4$ by means of the linear system $|3E_0 - \sum_{i=1}^{5} E_i|$ and $\deg(X) = 4$.
- Castelnuovo surface: $X = Bl_{\{p_1,\ldots,p_8\}}(\mathbb{P}^2)$ embedded in $\mathbb{P}^4$ by means of the linear system $|4E_0 - 2E_1 - \sum_{i=2}^{8} E_i|$ and $\deg(X) = 5$.
- Bordiga surface: $X = Bl_{\{p_1,\ldots,p_{10}\}}(\mathbb{P}^2)$ embedded in $\mathbb{P}^4$ by means of the linear system $|4E_0 - \sum_{i=1}^{10} E_i|$ and $\deg(X) = 6$.

This comes from the Riemann-Roch Theorem and the fact that $X$ is rational and ACM, which together give that $\deg X \leq 6$. Then the result follows easily from [**61**].

The following easy lemma will simplify our computations:

LEMMA 8.8. *Let $C = (a; b_1, ..., b_s)$ be a minimal ACM curve of degree $d$ and genus $g$ on a smooth rational, non-degenerate ACM surface $X \subset \mathbb{P}^4$. It holds:*

*1.- If $X$ is a cubic scroll then $g = 0$ and $d \leq 4$.*

*2.- If $X$ is a Del Pezzo surface then $g = 0$ and $d \leq 4$.*

*3.- If $X$ is a Castelnuovo surface then $g \leq \max\{0, a - b_1 - 1\}$ and $d \leq 11$.*

*4.- If $X$ is a Bordiga surface then $g \leq \max\{0, a - 2\}$ and $d \leq 15$.*

PROOF. 1.- On a cubic scroll $X$ we have $H + E_0 = -K$. Hence, the exact cohomology sequence associated to the exact sequence:

$$0 \longrightarrow \mathcal{O}_X(-C) \longrightarrow \mathcal{O}_X \longrightarrow \mathcal{O}_C \longrightarrow 0$$

gives us

$$g(C) = h^1(\mathcal{O}_C) = h^2(\mathcal{O}_X(-C)) = h^0(\mathcal{O}_X(C + K)) = h^0(\mathcal{O}_X(C - H - E_0)) = 0$$

where the last equality is a consequence of the minimality of $C$. Moreover, since $C$ is linearly normal, we obtain $\deg(C) - 4 \leq g(C) = 0$ which gives what we want.

2.- On a Del Pezzo surface $X$ we have $H = -K$. Using the above exact sequence we get $g(C) = h^1(\mathcal{O}_C) = h^2(\mathcal{O}_X(-C)) = h^0(\mathcal{O}_X(C+K)) = h^0(\mathcal{O}_X(C-H)) = 0$ where the last equality is a consequence of the minimality of $C$. Moreover, since $C$ is linearly normal, we obtain $\deg(C) - 4 \leq g(C) = 0$ which gives what we want.

3.- On a Castelnuovo surface $X$ we have $E_0 - E_1 = H + K$. Using the above exact sequence we get $g(C) = h^1(\mathcal{O}_C) = h^2(\mathcal{O}_X(-C)) = h^0(\mathcal{O}_X(C+K))$. The exact cohomology sequence associated to the exact sequence

$$0 \longrightarrow \mathcal{O}_X(C-H) \longrightarrow \mathcal{O}_X(C+K) \longrightarrow \mathcal{O}_{E_0-E_1}(C+K) \longrightarrow 0$$

together with the minimality of $C$ gives us

$$g = h^0(\mathcal{O}_X(C+K)) \leq h^0(\mathcal{O}_{E_0-E_1}(C+K)) = \max\{0, a - b_1 - 1\}.$$

Moreover, since $C$ is linearly normal, we obtain $\deg(C) - 4 \leq g(C)$. Combining this with Claim 1 of the proof of Proposition 8.4, we get what we want.

4.- On a Bordiga surface $X$ we have $H = E_0 - K$. Using the above exact sequence we get

$$g(C) = h^1(\mathcal{O}_C) = h^2(\mathcal{O}_X(-C)) = h^0(\mathcal{O}_X(C+K)) = h^0(\mathcal{O}_X(C-H+E_0)).$$

The exact cohomology sequence associated to the exact sequence:

$$0 \longrightarrow \mathcal{O}_X \longrightarrow \mathcal{O}_X(E_0) \longrightarrow \mathcal{O}_{E_0}(E_0) \longrightarrow 0$$

together with the minimality of $C$ gives us

$$g = h^0(\mathcal{O}_X(C-H+E_0)) \leq h^0(\mathcal{O}_{E_0}(C+K)) = \max\{0, a-2\}.$$

Moreover, since $C$ is linearly normal, we obtain $\deg(C) - 4 \leq g(C)$ which gives what we want. □

We say that a smooth rational ACM surface is *general* if it is obtained by blowing up points of $\mathbb{P}^2$ in general position. Now, using Claim 1 of the proof of Proposition 8.4 and Lemma 8.8 we will characterize ACM curves on a general smooth ACM rational surface and describe the corresponding minimal curves.

CASE 1: Cubic surface. A curve $C = (a; b) \in \operatorname{Pic}(X)$ is ACM if and only if is one of the following:

(i) $(2\nu; \nu - 1)$. The corresponding minimal ACM curve has $(d, g) = (1, 0)$,
(ii) $(2\nu; \nu)$,
(iii) $(2\nu + 1; \nu)$. The corresponding minimal ACM curve has $(d, g) = (2, 0)$,
(iv) $(2\nu + 1; \nu + 1)$. The corresponding minimal ACM curve has $(d, g) = (1, 0)$.

CASE 2: General Del Pezzo surface. A curve $C = (a; b_1, ..., b_5) \in \operatorname{Pic}(X)$ with $b_1 \geq ... \geq b_5$ is ACM if and only if is one of the following:

(i) $(3\nu; \nu^4, \nu - 1)$. The minimal ACM curve $(\nu = 0)$ has $(d, g) = (1, 0)$,
(ii) $(3\nu; \nu^5)$,
(iii) $(3\nu; \nu + 1, \nu^4)$. The minimal ACM curve $(\nu = 1)$ has $(d, g) = (3, 0)$,
(iv) $(3\nu; \nu + 1, \nu^3, \nu - 1)$. The minimal ACM curve $(\nu = 1)$ has $(d, g) = (4, 0)$,
(v) $(3\nu + 1; \nu^5)$. The minimal ACM curve $(\nu = 0)$ has $(d, g) = (3, 0)$,
(vi) $(3\nu + 1; \nu + 1, \nu^4)$. The minimal ACM curve $(\nu = 0)$ has $(d, g) = (2, 0)$,
(vii) $(3\nu + 1; \nu + 1^2, \nu^3)$. The minimal ACM curve $(\nu = 0)$ has $(d, g) = (1, 0)$,
(viii) $(3\nu + 1; \nu + 1^3, \nu^2)$. The minimal ACM curve $(\nu = 1)$ has $(d, g) = (4, 0)$,
(ix) $(3\nu + 2; \nu + 1^2, \nu^3)$. The minimal ACM curve $(\nu = 0)$ has $(d, g) = (4, 0)$,
(x) $(3\nu + 2; \nu + 1^3, \nu^2)$. The minimal ACM curve $(\nu = 0)$ has $(d, g) = (3, 0)$,

(xi) $(3\nu+2;\nu+1^4,\nu)$. The minimal ACM curve ($\nu=0$) has $(d,g)=(2,0)$,
(xii) $(3\nu+2;\nu+1^5)$. The minimal ACM curve ($\nu=0$) has $(d,g)=(1,0)$.

CASE 3: General Castelnuovo surface. A curve $C=(a;b_1,...,b_8)\in \mathrm{Pic}(X)$ with $b_2\geq ...\geq b_8$ is ACM if and only if is one of the following:

(i) $(4\nu;2\nu,\nu^6,\nu-1)$. The minimal ACM curve ($\nu=0$) has $(d,g)=(1,0)$,
(ii) $(4\nu;2\nu-1,\nu^7)$. The minimal ACM curve ($\nu=0$) has $(d,g)=(2,0)$,
(iii) $(4\nu;2\nu,\nu^7)$,
(iv) $(4\nu;2\nu+1,\nu^7)$. The minimal ACM curve ($\nu=1$) has $(d,g)=(3,0)$,
(v) $(4\nu;2\nu+1,\nu^6,\nu-1)$. The minimal ACM curve ($\nu=1$) has $(d,g)=(4,0)$,
(vi) $(4\nu;2\nu,\nu+1,\nu^6)$. The minimal ACM curve ($\nu=1$) has $(d,g)=(4,1)$,
(vii) $(4\nu;2\nu,\nu+1,\nu^5,\nu-1)$. The minimal ACM curve ($\nu=1$) has $(d,g)=(5,1)$,
(viii) $(4\nu;2\nu-1,\nu+1^2,\nu^5)$. The minimal ACM curve ($\nu=1$) has $(d,g)=(5,1)$,
(ix) $(4\nu;2\nu-1,\nu+1^3,\nu^4)$. The minimal ACM curve ($\nu=1$) has $(d,g)=(4,0)$,
(x) $(4\nu;2\nu-1,\nu+1,\nu^6)$. The minimal ACM curve ($\nu=1$) has $(d,g)=(6,2)$,
(xi) $(4\nu;2\nu,\nu+1^2,\nu^5)$. The minimal ACM curve ($\nu=1$) has $(d,g)=(3,0)$,
(xii) $(4\nu;2\nu,\nu+1^2,\nu^4,\nu-1)$. The minimal ACM curve ($\nu=1$) has $(d,g)=(4,0)$,
(xiii) $(4\nu+1;2\nu,\nu^7)$. The minimal ACM curve ($\nu=0$) has $(d,g)=(4,0)$,
(xiv) $(4\nu+1;2\nu+1,\nu^7)$. The minimal ACM curve ($\nu=0$) has $(d,g)=(2,0)$,
(xv) $(4\nu+1;2\nu,\nu+1,\nu^6)$. The minimal ACM curve ($\nu=0$) has $(d,g)=(3,0)$,
(xvi) $(4\nu+1;2\nu+1,\nu+1,\nu^6)$. The minimal ACM curve ($\nu=0$) has $(d,g)=(1,0)$,
(xvii) $(4\nu+1;2\nu,\nu+1^2,\nu^5)$. The minimal ACM curve ($\nu=0$) has $(d,g)=(2,0)$,
(xviii) $(4\nu+1;2\nu+1,\nu+1^2,\nu^5)$. The minimal ACM curve ($\nu=1$) has $(d,g)=(5,1)$,
(xix) $(4\nu+1;2\nu,\nu+1^3,\nu^4)$. The minimal ACM curve ($\nu=1$) has $(d,g)=(6,2)$,
(xx) $(4\nu+1;2\nu,\nu+1^4,\nu^3)$. The minimal ACM curve ($\nu=1$) has $(d,g)=(5,1)$,
(xxi) $(4\nu+1;2\nu,\nu+1^5,\nu^2)$. The minimal ACM curve ($\nu=1$) has $(d,g)=(4,0)$,
(xxii) $(4\nu+1;2\nu+1,\nu+1^3,\nu^4)$. The minimal ACM curve ($\nu=1$) has $(d,g)=(4,0)$,
(xxiii) $(4\nu+2;2\nu+1,\nu+1^2,\nu^5)$. The minimal ACM curve ($\nu=0$) has $(d,g)=(4,0)$,
(xxiv) $(4\nu+2;2\nu,\nu+1^4,\nu^3)$. The minimal ACM curve ($\nu=0$) has $(d,g)=(4,0)$,
(xxv) $(4\nu+2;2\nu+1,\nu+1^3,\nu^4)$. The minimal ACM curve ($\nu=0$) has $(d,g)=(3,0)$,
(xxvi) $(4\nu+2;2\nu,\nu+1^5,\nu^2)$. The minimal ACM curve ($\nu=0$) has $(d,g)=(3,0)$,
(xxvii) $(4\nu+2;2\nu+1,\nu+1^4,\nu^3)$. The minimal ACM curve ($\nu=0$) has $(d,g)=(2,0)$,
(xxviii) $(4\nu+2;2\nu+1,\nu+1^5,\nu^2)$. The minimal ACM curve ($\nu=1$) has $(d,g)=(6,2)$,
(xxix) $(4\nu+2;2\nu,\nu+1^6,\nu)$. The minimal ACM curve ($\nu=1$) has $(d,g)=(7,3)$,
(xxx) $(4\nu+2;2\nu+1,\nu+1^6,\nu)$. The minimal ACM curve ($\nu=1$) has $(d,g)=(5,1)$,
(xxxi) $(4\nu+2;2\nu,\nu+1^7)$. The minimal ACM curve ($\nu=1$) has $(d,g)=(6,2)$,
(xxxii) $(4\nu+2;2\nu+1,\nu+1^7)$. The minimal ACM curve ($\nu=1$) has $(d,g)=(4,0)$,
(xxxiii) $(4\nu+3;2\nu+1,\nu+1^5,\nu^2)$. The minimal ACM curve ($\nu=0$) has $(d,g)=(5,1)$,
(xxxiv) $(4\nu+3;2\nu+1,\nu+1^6,\nu)$. The minimal ACM curve ($\nu=0$) has $(d,g)=(4,1)$,
(xxxv) $(4\nu+3;2\nu,\nu+1^7)$. The minimal ACM curve ($\nu=0$) has $(d,g)=(5,1)$,
(xxxvi) $(4\nu+3;2\nu+1,\nu+1^7)$. The minimal ACM curve ($\nu=0$) has $(d,g)=(3,1)$,
(xxxvii) $(4\nu+3;2\nu+2,\nu+1^4,\nu^3)$. The minimal ACM curve ($\nu=0$) has $(d,g)=(4,0)$,
(xxxviii) $(4\nu+3;2\nu+2,\nu+1^5,\nu^2)$. The minimal ACM curve ($\nu=0$) has $(d,g)=(3,0)$,
(xxxix) $(4\nu+3;2\nu+2,\nu+1^6,\nu)$. The minimal ACM curve ($\nu=0$) has $(d,g)=(2,0)$,
(xl) $(4\nu+3;2\nu,\nu+2,\nu+1^6)$. The minimal ACM curve ($\nu=0$) has $(d,g)=(4,0)$,
(xli) $(4\nu+3;2\nu+1,\nu+2,\nu+1^4,\nu^2)$. The minimal ACM curve ($\nu=0$) has $(d,g)=(4,0)$,

(xlii) $(4\nu+3; 2\nu+1, \nu+2, \nu+1^5, \nu)$. The minimal ACM curve ($\nu=0$) has $(d,g)=(3,0)$,
(xliii) $(4\nu+3; 2\nu+1, \nu+2, \nu+1^6)$. The minimal ACM curve ($\nu=1$) has $(d,g)=(7,3)$,
(xliv) $(4\nu+3; 2\nu+2, \nu+1^7)$. The minimal ACM curve ($\nu=1$) has $(d,g)=(6,2)$.

CASE 4: General Bordiga surface. A curve $C=(a; b_1, ..., b_{10}) \in \mathrm{Pic}(X)$ with $b_1 \geq ... \geq b_{10}$ is ACM if and only if is one of the following:

(i) $(4\nu; \nu^9, \nu-1)$. The minimal ACM curve ($\nu=0$) has $(d,g)=(1,0)$,
(ii) $(4\nu; \nu^{10})$,
(iii) $(4\nu; \nu+1, \nu^9)$. The minimal ACM curve ($\nu=1$) has $(d,g)=(5,2)$,
(iv) $(4\nu; \nu+1^2, \nu^8)$. The minimal ACM curve ($\nu=1$) has $(d,g)=(4,1)$,
(v) $(4\nu; \nu+1^3, \nu^7)$. The minimal ACM curve ($\nu=2$) has $(d,g)=(9,5)$,
(vi) $(4\nu; \nu+1, \nu^8, \nu-1)$. The minimal ACM curve ($\nu=1$) has $(d,g)=(6,2)$,
(vii) $(4\nu; \nu+1^2, \nu^7, \nu-1)$. The minimal ACM curve ($\nu=1$) has $(d,g)=(5,1)$,
(viii) $(4\nu; \nu+1^3, \nu^6, \nu-1)$. The minimal ACM curve ($\nu=2$) has $(d,g)=(10,6)$,
(ix) $(4\nu+1; \nu^{10})$. The minimal ACM curve ($\nu=0$) has $(d,g)=(4,0)$,
(x) $(4\nu+1; \nu+1, \nu^9)$. The minimal ACM curve ($\nu=0$) has $(d,g)=(3,0)$,
(xi) $(4\nu+1; \nu+1^2, \nu^8)$. The minimal ACM curve ($\nu=0$) has $(d,g)=(2,0)$,
(xii) $(4\nu+1; \nu+1^3, \nu^7)$. The minimal ACM curve ($\nu=1$) has $(d,g)=(7,3)$,
(xiii) $(4\nu+1; \nu+1^4, \nu^6)$. The minimal ACM curve ($\nu=1$) has $(d,g)=(6,2)$,
(xiv) $(4\nu+1; \nu+1^5, \nu^5)$. The minimal ACM curve ($\nu=1$) has $(d,g)=(5,1)$,
(xv) $(4\nu+1; \nu+1^6, \nu^4)$. The minimal ACM curve ($\nu=2$) has $(d,g)=(10,6)$,
(xvi) $(4\nu+2; \nu+1^4, \nu^6)$. The minimal ACM curve ($\nu=0$) has $(d,g)=(4,0)$,
(xvii) $(4\nu+2; \nu+1^5, \nu^5)$. The minimal ACM curve ($\nu=0$) has $(d,g)=(3,0)$,
(xviii) $(4\nu+2; \nu+1^6, \nu^4)$. The minimal ACM curve ($\nu=1$) has $(d,g)=(8,4)$,
(xix) $(4\nu+2; \nu+1^7, \nu^3)$. The minimal ACM curve ($\nu=1$) has $(d,g)=(7,3)$,
(xx) $(4\nu+2; \nu+1^8, \nu^2)$. The minimal ACM curve ($\nu=1$) has $(d,g)=(6,2)$,
(xxi) $(4\nu+2; \nu+1^9, \nu)$. The minimal ACM curve ($\nu=2$) has $(d,g)=(11,8)$,
(xxii) $(4\nu+2; \nu+1^{10})$. The minimal ACM curve ($\nu=2$) has $(d,g)=(10,6)$,
(xxiii) $(4\nu+3; \nu+1^7, \nu^3)$. The minimal ACM curve ($\nu=0$) has $(d,g)=(5,1)$,
(xxiv) $(4\nu+3; \nu+1^8, \nu^2)$. The minimal ACM curve ($\nu=0$) has $(d,g)=(4,1)$,
(xxv) $(4\nu+3; \nu+1^9, \nu)$. The minimal ACM curve ($\nu=0$) has $(d,g)=(3,1)$,
(xxvi) $(4\nu+3; \nu+1^{10})$. The minimal ACM curve ($\nu=1$) has $(d,g)=(8,5)$,
(xxvii) $(4\nu+3; \nu+2, \nu+1^6, \nu^3)$. The minimal ACM curve ($\nu=0$) has $(d,g)=(4,0)$,
(xxviii) $(4\nu+3; \nu+2, \nu+1^7, \nu^2)$. The minimal ACM curve ($\nu=1$) has $(d,g)=(9,5)$,
(xxix) $(4\nu+3; \nu+2, \nu+1^8, \nu)$. The minimal ACM curve ($\nu=1$) has $(d,g)=(8,4)$,
(xxx) $(4\nu+3; \nu+2, \nu+1^9)$. The minimal ACM curve ($\nu=1$) has $(d,g)=(7,3)$.

REMARK 8.9. Using the same arguments we use to produce the above list, we can write down the list of ACM minimal curves lying on a smooth ACM rational surface, not necessarily general, and it turns out that the list in the smooth "non-general" case is a subset of the list produced above, decreasing the value of $\nu$ if necessary.

We believe that Corollary 8.10 below holds for *any* smooth rational ACM surface, without needing to add the word "general." However, our proof uses the generality in a strong way, and while we believe that the same approach could yield a proof of the "non-general" case, it seems impractical to check all the possible configurations of points that could arise, and all the ensuing cases.

Now, we are ready to prove that there is only one G-liaison class containing ACM curves $C \subset \mathbb{P}^4$ lying on a "general" smooth rational ACM surface $X$ of $\mathbb{P}^4$.

COROLLARY 8.10. *All ACM curves $C \subset \mathbb{P}^4$ lying on a "general" smooth, rational ACM surface $X \subset \mathbb{P}^4$ are glicci.*

PROOF. By Lemma 8.3 we only need to check that all minimal ACM curves $C$ lying on a "general" smooth, rational, ACM surface $X \subset \mathbb{P}^4$ are glicci. We will assume that $X$ is a Bordiga (resp. Castelnuovo) surface. The other cases are left to the reader.

In the above list if $d \leq 3$ or $(d,g) = (4,1)$ or $(5,2)$, then the curve $C$ is degenerate and it is well known that ACM curves in $\mathbb{P}^3$ are licci, hence glicci. On the other hand, in the above list all minimal ACM curves with $d \geq 8$ (resp. $d \geq 5$) are G-linked to an ACM curve with $d < 8$ (resp. $d < 4$) via the Gorenstein divisor $tH - K$ for a suitable $t \in \mathbb{Z}$. So, we only have to discuss the cases $(d,g) = (4,0)$, $(5,1)$, $(6,2)$ and $(7,3)$ (resp. $(4,0)$).

We first remark that by Corollary 5.14, it is enough to prove the glicciness of the general element of a linear system in order to prove it for *every* element.

Checking the list for the "general" Bordiga surface, one sees that each of the linear systems corresponding to $(d,g) = (4,0)$, $(5,1)$, $(6,2)$ and $(7,3)$ possesses a general element that is integral. Most of these linear systems move in at least a pencil. Only (xiv) and (xxvii) are effective but do not move in a linear system. But in these latter cases, since the surface is general we can choose the points to blow up in a general enough way to get the curve to be integral: in (xiv) there is exactly one plane quintic double at 5 general points and passing through 5 more general points, and this is integral; similarly in (xxvii) there is a unique plane cubic double at one point and passing through six of the others, and it is also integral.

Checking the list for the "general" Castelnuovo surface, one sees that there are 12 linear systems with $(d,g) = (4,0)$. Six of them are G-linked (via $H - K$) to the other six. Hence without loss of generality we may restrict our attention to the following cases:

| | |
|---|---|
| (xiii) | $(1; 0, 0^7)$ |
| (xxiii) | $(2; 1, 1^2, 0^5)$ |
| (xxiv) | $(2; 0, 1^4, 0^3)$ |
| (xxxvii) | $(3; 2, 1^4, 0^3)$ |
| (xl) | $(3; 0, 2, 1^6)$ |
| (xli) | $(3; 1, 2, 1^4, 0^2)$ |

Exactly as above, we can show that for the "general" Castelnuovo surface each of these has an integral general element. (Note that (xl) corresponds to a unique curve (effective divisor), but it is integral by the general choice of the points of $\mathbb{P}^2$ which were blown up.)

Now, an integral ACM curve $C$ with $(d,g) = (4,0)$ is in fact smooth. The union of $C$ with a general secant line (whether or not it lies on the surface) is arithmetically Gorenstein, hence $C$ is glicci.

An integral ACM curve with $(d,g) = (5,1)$ is arithmetically Gorenstein since it is non-degenerate and so its general hyperplane section consists of 5 points in $\mathbb{P}^3$ with symmetric $h$-vector and uniform position, hence the Cayley-Bacharach property and so the main theorem of [**17**] applies.

We now turn to ACM curves with $(d, g) = (6, 2)$ and $(7, 3)$ on a "general" Bordiga surface, and as above we can assume that the curves are integral. One computes that $h^0(\mathcal{I}_C(2)) = 4$ or 3, respectively. Choose three general (hence independent) quadrics containing $C$. If they form a complete intersection then the residual to $C$ is known to be glicci (even licci). So we analyze the other possibilities.

Certainly the first two quadrics do not contain a common factor since $C$ is integral, so they define a complete intersection surface $T$. If $T$ is integral then the three quadrics define a complete intersection curve and we are done. If all components of $T$ have degree $\leq 2$ then $C$ must lie on one of these components and hence is degenerate, and thus licci.

The only remaining possibility is that $T$ is the union of an irreducible cubic surface $X$ and a plane. But then the cubic $X$ is ACM (by liaison). If $X$ is smooth, this is the cubic scroll mentioned above. The other possibility is that the cubic surface $X$ is a cone over a twisted cubic curve $Y$ in $\mathbb{P}^3$. Then it follows from [**31**], II, Exercise 6.3 that the divisor class group $\mathrm{Cl}\, X \cong \mathrm{Cl}\, Y \cong \mathbb{Z}$ (since $Y \cong \mathbb{P}^1$). In particular, as in the case where $X$ is smooth, every ACM curve on $X$ is linearly equivalent to a curve of the form $C_0 + tH$ where $\deg C_0 \leq 3$ and $C_0$ is ACM. (If $\deg C_0 = 3$, it is a hyperplane section.) But then in each case $C_0$ is degenerate, so we are done.

The analysis of the cases $(d, g) = (5, 1), (6, 2)$ and $(7, 3)$ on a "general" Bordiga surface were first verified using the computer program Macaulay [**5**]. □

REMARK 8.11. A consequence of the above proof is that all *integral* ACM curves in $\mathbb{P}^4$ with $(d, g) = (4, 0), (5, 1), (6, 2)$ and $(7, 3)$ are glicci. The proof of this fact did not rely on the existence of the rational surface.

Using the classification of minimal ACM curves, $C$, on a smooth rational nondegenerate ACM surface, $S \subset \mathbb{P}^4$, we can also show that many ACM curves on a cubic scroll, Castelnuovo surface or Bordiga surface are not licci. More generally, we show that $C_\nu \in |C + \nu H|$ and $C_{\nu'} \in |C + \nu' H|$ belong to different CI-liaison classes for $\nu > \nu' \gg 0$. Indeed this follows immediately from the complex (7.2) and Corollary 7.6 if we can show that for some integer $\nu$

$$h^0(\omega_S(5) \otimes \mathcal{I}_{C|S}(\nu)) > \sum_i h^0(\mathcal{I}_{C|S}(q_i + \nu)). \tag{8.3}$$

The first step is to find an appropriate lower bound for $h^0(\omega_S(5) \otimes \mathcal{I}_{C|S}(\nu))$.

LEMMA 8.12. *Let $C$ be a minimal ACM curve on a smooth nondegenerate ACM surface $S$. Let $H$ denote the hyperplane divisor and $K$ the canonical divisor. Assume $C \notin |H|$. Then*

$$\begin{aligned} h^0(\omega_S(n+1) \otimes \mathcal{I}_{C|S}) \;\geq\; & \tfrac{1}{2}(2n+1)H(H+K) - C(H+K) \\ & + h^0(\mathcal{I}_{C|S}(n)) + h^1(\mathcal{O}_C(n)) + h^2(\mathcal{O}_S(n)) \quad \text{for } n \geq 0. \end{aligned}$$

PROOF. Since $C$ is minimal and $C \notin |H|$, we get

$$h^2(\omega_S(n+1) \otimes \mathcal{I}_{C|S}) = h^0(\mathcal{O}_S(C - (n+1)H)) = 0 \text{ for } n \geq 0.$$

By the Riemann-Roch Theorem, the adjunction formula $C^2 + C \cdot K = 2g - 2$ and the equality $C(K + H) = C \cdot K + d$, we have:

$$\begin{aligned} h^0(\omega_S(n+1) \otimes \mathcal{I}_{C|S}) &\geq \chi(\mathcal{O}_S(K + (n+1)H - C)) \\ &= \tfrac{1}{2}(K + (n+1)H - C) \cdot ((n+1)H - C) + 1 + h^2(\mathcal{O}_S) \\ &= \tfrac{1}{2}((n+1)H \cdot K + (n+1)^2 H^2) \\ &\qquad -nd + g - C(K+H) + h^2(\mathcal{O}_S). \end{aligned}$$

Using the Riemann-Roch Theorem on $C$ we get

$$\begin{aligned} nd - g &= -1 + \chi(\mathcal{O}_C(n)) \\ &= -1 + h^0(\mathcal{O}_C(n)) - h^1(\mathcal{O}_C(n)) \\ &= -1 - h^1(\mathcal{O}_C(n)) + h^0(\mathcal{O}_S(n)) - h^0(\mathcal{I}_{C|S}(n)) \end{aligned}$$

while the Riemann-Roch Theorem for $S$ implies

$$h^0(\mathcal{O}_S(n)) + h^2(\mathcal{O}_S(n)) = \frac{1}{2}(n^2H^2 - nH \cdot K) + 1 + h^2(\mathcal{O}_S).$$

Combining we obtain

$$\begin{aligned} h^0(\omega_S(n+1) \otimes \mathcal{I}_{C|S}) &\geq \tfrac{1}{2}((n+1)H \cdot K + (n+1)^2H^2) - C(K+H) \\ &\quad + h^1(\mathcal{O}_C(n)) + h^0(\mathcal{I}_{C|S}(n)) + h^2(\mathcal{O}_S(n)) \\ &\quad - \tfrac{1}{2}(n^2H^2 - nH \cdot K) \\ &= \tfrac{1}{2}(2n+1)H(H+K) - C(H+K) \\ &\quad + h^0(\mathcal{I}_{C|S}(n)) + h^1(\mathcal{O}_C(n)) + h^2(\mathcal{O}_S(n)) \text{ for } n \geq 0 \end{aligned}$$

which proves the lemma. □

In particular, we have

PROPOSITION 8.13. *Let $C$ be a minimal ACM curve, $C \notin |H|$, on a smooth rational non-degenerate ACM surface $S \subset \mathbb{P}^4$. $C$ will be denoted by the vector $(a; b_1, b_2, \dots)$ in the usual basis. If*

(i) *$S$ is a cubic scroll and $a > 2n + 1 - h^0(\mathcal{I}_{C|S}(n))$ for some $n$ with $0 \leq n \leq s(C|S)$ or*

(ii) *$S$ is a Castelnuovo surface and $a - b_1 < 2n+1-h^0(\mathcal{I}_{C|S}(n))$ for some $n \geq 0$ or*

(iii) *$S$ is a Bordiga surface and $a < 4n + 2 - 2h^0(\mathcal{I}_{C|S}(n))$ for some $n \geq 0$*

*then $C_\nu$ and $C_{\nu'}$ belong to different CI-liaison classes for $\nu > \nu' \gg 0$.*

PROOF. (i) In this case $K = (-3; -1), H = (2; 1), H(H + K) = -2$ and the minimal resolution of $I_S$ is:

$$0 \to R(-3)^{\oplus 2} \to R(-2)^{\oplus 3} \to I_S \to 0.$$

By (8.3) it suffices to show $h^0(\omega_S(n+1) \otimes \mathcal{I}_{C|S}) > 0$ for some $n$ with $0 \leq n \leq s(C|S)$. By Lemma 8.12, we have

$$h^0(\omega_S(n+1) \otimes \mathcal{I}_{C|S}) \geq \frac{1}{2}(2n+1)(-2) + a + h^0(\mathcal{I}_{C|S}(n))$$

and we conclude by the assumption of (i) and Corollary 7.6.

(ii) In this case $K = (-3; -1^8), H = (4; 2, 1^7), H(H+K) = 2$ and $C(H+K) = a - b_1$ and the minimal resolution of $I_S$ is:

$$0 \to R(-4)^{\oplus 2} \to R(-3)^{\oplus 2} \oplus R(-2) \to I_S \to 0.$$

By (8.3) it suffices to show $h^0(\omega_S(n+1) \otimes \mathcal{I}_{C|S}) > 2h^0(\mathcal{I}_{C|S}(n))$ for some $n \geq 0$. By Lemma 8.12 we have

$$2n + 1 - (a - b_1) + h^0(\mathcal{I}_{C|S}(n)) > 2h^0(\mathcal{I}_{C|S}(n))$$

and we conclude by the assumption of (ii) and Corollary 7.6.

(iii) In this case $K = (-3; -1^{10}), H = (4; 1^{10}), H(H + K) = 4, C(H + K) = a$ and the minimal resolution of $I_S$ is:

$$0 \to R(-4)^{\oplus 3} \to R(-3)^{\oplus 4} \to I_S \to 0.$$

Again we conclude by (8.3), Lemma 8.12, the assumption (iii) and Corollary 7.6. □

REMARK 8.14. (i) On a cubic scroll the family (iv) satisfies Lemma 8.13 (i) for $n = 1$ so it gives rise to an infinite number of CI-liaison classes containing ACM curves.

(ii) On a Castelnuovo surface the families (i), (xiv), (xvi) (resp. (iv), (v), (vii), (xii), (xiii), (xiv), (xv), (xviii), (xxii), (xxiii), (xxiv), (xxv), (xxxiii), (xxxvii), (xxxviii), (xli)) satisfy Lemma 8.13 (ii) for $n = 0$ (resp. $n = 1$).

(iii) On a Bordiga surface the families (i), (ix), (x), (xi) (resp. (vi), (vii), (viii), (xii)-(xvii), (xxii), (xxiv), (xxvii)) satisfy Lemma 8.13 (iii) for $n = 0$ (resp. $n = 1$). The family (xxviii) satisfies Lemma 8.13 (iii) for $n = 2$.

CHAPTER 9

# Unobstructedness and dimension of families of subschemes

In this chapter we consider objects $C = \text{Proj}(A) \subset S = \text{Proj}(B) \subset \mathbb{P}^N$ of equidimensional, locally Cohen-Macaulay closed subschemes, *where $C$ is a Cartier divisor on $S$.* In most cases we suppose the local algebra cohomology group $H^2(R, B, B)^\sim$ of $S$ vanishes (e.g. $S$ has the property $G_0$ and is locally licci; cf. Proposition 6.17). Since we often focus on subschemes $C \subset \mathbb{P}^N$ of codimension 3, the mentioned assumptions on $S$ are weak (i.e. they hold if S has $G_0$). For subschemes of higher codimension we can avoid the vanishing of $H^2(R, B, B)^\sim$ (Remark 9.5). Finally $C$ and $S$ are ACM each time we explicitly call them ACM schemes.

Our purpose is to prove the unobstructedness of most of the subschemes $C_t$ we obtain by adding a hypersurface section of $S \subset \mathbb{P}^N$ of degree $t$ to $C \subset S$, and, moreover, to find manageable formulas for the dimension of the corresponding families of their Hilbert schemes. Our results also allow us to find the dimension of the Hilbert scheme for some classes of ACM curves in $\mathbb{P}^4$ and to show the unobstructedness of many classes of determinantal subschemes in $\mathbb{P}^N$ in the next chapter. For a general reference for obstruction theory, see [**47**].

As we have seen, such subschemes (obtained by varying the parameters in the matrix or by varying $t$) usually belong to different CI-liaison classes and to the same G-liaison class. Unobstructedness (even strongly unobstructedness) of ACM schemes is preserved under CI-linkage, cf. [**38**], Proposition 3.12 and [**13**], and one might wonder whether unobstructedness is preserved under G-liaison as well. This is, however, not true in general. Indeed we will see in Remark 9.12 that there exist obstructed standard determinantal curves in $\mathbb{P}^4$. Since any such determinantal curve is in the G-liaison class of a complete intersection by Theorem 3.6, this will provide a counterexample because complete intersections are necessarily unobstructed. On the other hand, we know that the process of adding hypersurface sections to a divisor $C$ on an ACM subscheme $S$ is a certain G-biliaison (Proposition 5.10 and Corollary 5.14). Our results on the unobstructedness of $C_t$ therefore show that unobstructedness is preserved under G-biliaison under certain conditions.

Recall that J. Herzog [**32**] calls $A = R/I$ *strongly unobstructed* if $A$ is generically a complete intersection and $A$ and $I/I^2 \otimes_A K_A$ are Cohen-Macaulay. Indeed, for a generic complete intersection of dimension $\dim A = n + 1$, we have by duality

$$H^n_{\mathfrak{m}}(K_A \otimes_R I)^\vee \cong \text{Ext}^1_A(I/I^2, A) \cong H^2(R, A, A) \tag{9.1}$$

which vanishes by Cohen-Macaulayness. A closed subscheme $C$ of $\mathbb{P}^N$ is called *unobstructed* if $\text{Hilb}^p := \text{Hilb}^p(\mathbb{P}^N)$ is smooth at $(C \subset \mathbb{P}^N)$. Especially when $\dim C = 0$, we also consider the strata $\text{GradAlg}^H = \text{GradAlg}^H(\mathbb{P}^N)$ of $\text{Hilb}^p$, where we deform

$C \subset \mathbb{P}^N$ with constant Hilbert function $H = H_A = H_C$. $\text{GradAlg}^H$ allows a natural scheme structure [**41**] whose tangent (resp. "obstruction") space is ${}_0H^1(R, A, A)$ (resp. ${}_0H^2(R, A, A)$), i.e. it is given by deforming $R \to A$ as a graded algebra. $C$ will be called *unobstructed in* $\text{GradAlg}^H$ (or $H$- *unobstructed*) if $\text{GradAlg}^H$ is smooth at $(C \subset \mathbb{P}^N)$. By [**40**], Remark 3.7, $\text{GradAlg}^H \cong \text{Hilb}^p$ at $(C \subset \mathbb{P}^N)$ under the assumption ${}_0\text{Hom}_R(I, H^2_{\mathfrak{m}}(I)) = 0$. The two concepts of unobstructedness of $C$ are therefore equivalent if $C$ is an ACM subscheme of positive dimension. So, for a positive dimensional ACM scheme, the vanishing of ${}_0H^2(R, A, A)$ is sufficient (but not necessary by Example 7.5 and Proposition 10.7) for unobstructedness, and, in a sense, we therefore see that our definition of unobstructedness differs from that in [**32**], which requires the whole group $H^2(R, A, A)$ to vanish (to have the unobstructedness of $A$).

By (9.1) and the considerations above, we remark that the vanishing of the liaison-invariant ${}_0H^n_{\mathfrak{m}}(K_A \otimes_R I)$ suffices for a *generic complete intersection* ACM scheme $C$ to be unobstructed. For *every* ACM scheme $C$ of codimension 3, Remark 6.16 implies the following: If the liaison-invariant ${}_0H^0_{\mathfrak{m}}(K_A(N+1) \otimes_R I) = 0$ (note that ${}_0H^0_{\mathfrak{m}}(K_A(N+1) \otimes_R I) \hookrightarrow H^1(\mathcal{N}_C)$ if $n > 0$ by Remark 6.3) and $H^2(R, A, A)^\sim = 0$ (cf. Proposition 6.17), then $C$ is unobstructed.

In codimension 3, Corollary 7.4 can be used as a criterion for unobstructedness. It is based on $L^0(C)_0 = 0$, i.e. it requires the right map in (7.2) to be surjective. In what follows, we will prove more general results, which in codimension 3 are based on the exactness of the complex (7.2) "in the middle."

First we consider the problem of deciding when an arbitrary $C_t \in |C + tH|$ (given by some non-zero divisor of $H^0(\mathcal{O}_S(C)(t))$) is unobstructed, provided (if necessary) $C$ is unobstructed. As a special case we will see when any member of $|C|$ is unobstructed if one member is. Similar problems for the case of smooth curves on a smooth surface were considered in [**39**], Section 2. It turns out that the assumption "$C$ unobstructed" is not too important because the assumption we really need is "$D(p, q)$ is smooth at $(C \subset S \subset \mathbb{P}^N)$" or "the relative Picard scheme Pic is smooth at $(\mathcal{O}_S(C), S)$." Here the relative Picard scheme $\text{Pic} = \text{Pic}^q$ parameterizes pairs $(L, S)$ of invertible sheaves on a (variable) subscheme $S \subset \mathbb{P}^N$ of Hilbert polynomial $q$, and as on page 36 we let $D(p, q)$ be the Hilbert flag scheme, which parameterizes "pairs" $C \subset S$ of equidimensional, locally Cohen-Macaulay closed subschemes of $\mathbb{P}^N$ with Hilbert polynomial $p$ and $q$ respectively.

REMARK 9.1. The local relative Picard functor is defined by deforming a given pair $(L, S)$ in the category of local artinian $k$-algebras with residue field $k$. Since we will not care about the existence of Pic, nor of the existence of the rational map $D(p, q) \to \text{Pic}$, essentially given by $(C \subset S \subset \mathbb{P}^N) \mapsto (\mathcal{O}_S(C), S)$, we will by definition let the phrase "Pic *is formally smooth at* $(L, S)$" mean that the local Picard functor at $(L, S)$ is formally smooth, and we let $\dim_{(L,S)} \text{Pic}$ be the dimension of the hull of the local functor [**47**]. Similarly, we say "$D(p, q) \to \text{Pic}$ is *formally smooth at* $(C \subset S \subset \mathbb{P}^N)$" if $C \subset S$ is Cartier and the corresponding morphism of the local deformation functors is formally smooth.

Note that the assumption $H^2(R, B, B)^\sim = 0$ (e.g. $G_0$ and locally licci) of Lemma 7.1 implies the exact sequences

$$0 \to \mathcal{N}_{C|S} \to \mathcal{N}_C \to \mathcal{N}_S \otimes \mathcal{O}_C \to 0$$

and

$$0 \to \mathcal{O}_S \to L \to \mathcal{N}_{C|S} \to 0,$$

where $L = \mathcal{O}_S(C)$. Their long exact sequences of cohomology define maps

$$H^0(\mathcal{N}_S) \to H^0(\mathcal{N}_S \otimes \mathcal{O}_C) \to H^1(\mathcal{N}_{C|S}) \to H^2(\mathcal{O}_S)$$

whose composition (resp. the composition $H^0(\mathcal{N}_S) \to H^0(\mathcal{N}_S \otimes \mathcal{O}_C) \to H^1(\mathcal{N}_{C|S})$) we call $\alpha_L$ (resp. $\alpha_{C|S}$). To handle the assumptions of Theorem 9.4 below, we need

PROPOSITION 9.2. *Let $(C \subset S \subset \mathbb{P}^N) \in D(p,q)$, $C$ a Cartier divisor on $S$, and suppose $S$ is unobstructed and satisfies $H^2(R,B,B)^\sim = 0$. Then the relative Picard scheme* Pic *is formally smooth at $(\mathcal{O}_S(C), S)$ if one of the following conditions holds;*

(i) *the map $\alpha_L : H^0(\mathcal{N}_S) \to H^2(\mathcal{O}_S)$ is surjective ;*
(ii) *$(C \subset S \subset \mathbb{P}^N)$ is a smooth point of $D(p,q)$ (true if $\alpha_{C|S}$ is surjective) and $H^1(\mathcal{O}_S(C)) = 0$;*
(iii) *$C$ is unobstructed, $H^1(\mathcal{O}_S(C)) = 0$ and $H^1(\mathcal{N}_S \otimes \mathcal{I}_{C|S}) \to H^1(\mathcal{N}_S)$ is injective.*

REMARK 9.3. The surjectivity of the map $\alpha_L$, together with the vanishing of $H^1(\mathcal{O}_S(C))$, may be seen as a generalization of the infinitesimal variant of [**43**], Section 2, for $(L,S)$ to be "general in the Noether-Lefschetz locus." In particular, if $\text{Pic}^q$ has different components, then we expect the general point $(L,S)$ of at least one of the components to have a surjective $\alpha_L$.

More importantly, there is a connection between $\operatorname{coker} \alpha_L$, or rather $\operatorname{coker} \alpha_{C|S}$, and the liaison-invariant group ${}_0H^0_{\mathfrak{m}}(K_A(N+1) \otimes_R I)$. Indeed, if we restrict to ACM subschemes $C \subset S$ in $\mathbb{P}^{n+3}$ of dimension $\dim C = n > 0$ where $S$ has $G_0$, then Proposition 7.2 (i) and Proposition 7.3 show there is an exact sequence

$$0 \to \operatorname{coker} \alpha_{C|S} \to {}_0H^0_{\mathfrak{m}}(K_A(n+4) \otimes_R I) \to L^0(C)_0 \to 0$$

provided ${}_0\operatorname{Hom}(I_{C|S}, H^{n+1}_{\mathfrak{m}}(A)) = 0$ (resp. $H^2(\mathcal{N}_{C|S}) = 0$) for $n = 1$ (resp. $n \geq 2$) and $H^0(\mathcal{N}_S) \to H^0(\mathcal{N}_S \otimes \mathcal{O}_C)$ is surjective.

*Proof of 9.2* We first prove (i). Since $H^2(\mathcal{O}_S)$ contains the obstruction $o(L)$ of deforming $L = \mathcal{O}_S(C)$ to a given deformation of $S$ by [**27**], the surjectivity of $\alpha_L$ implies the vanishing of any such obstruction after having changed the deformation of $S$ by a $\gamma$ of $H^0(\mathcal{N}_S)$ such that $\alpha_L(\gamma) = o(L)$; cf. [**42**], p. 324, which describes the tangent spaces of the deformations involved cohomologically. This proves (i).

We now prove (ii). First, by the arguments of (i), one may show that the surjectivity of $\alpha_{C|S}$ implies the smoothness of $D(p,q)$ at $(C \subset S \subset \mathbb{P}^N)$ (or one may use [**38**], (1.9) to see it). Moreover by [**42**], p. 324, or [**27**], Remark 4.5, the rational map $D(p,q) \to \text{Pic}$, essentially given by $(C \subset S \subset \mathbb{P}^N) \mapsto (\mathcal{O}_S(C), S)$ is formally smooth at $(C \subset S \subset \mathbb{P}^N)$ provided $H^1(\mathcal{O}_S(C)) = 0$, i.e. we get (ii).

Finally, we prove (iii). We have the following result ([**37**], Theorem 1.3.4, or [**38**], (1.15)): The first projection $pr_1 : D(p,q) \to \text{Hilb}^p$, given by

$$(C \subset S \subset \mathbb{P}^N) \mapsto (C \subset \mathbb{P}^N),$$

is smooth at $(C \subset S \subset \mathbb{P}^N)$ provided $S$ is unobstructed and $\operatorname{Hom}(\mathcal{I}_S, \mathcal{O}_S) \to \operatorname{Hom}(\mathcal{I}_S, \mathcal{O}_C)$ is surjective. To help understand this result we remark that, at $(C \subset S \subset \mathbb{P}^N)$, the tangent space of $D(p,q)$ (resp. tangent map of $pr_1 : D(p,q) \to \text{Hilb}^p$) is given by taking global sections of the corresponding sheaf (resp. morphism of sheaves) in the diagram (6.1) with $C \subset S$ instead of $X \subset Y$. Since

$\operatorname{Hom}(\mathcal{I}_S, \mathcal{O}_S) \to \operatorname{Hom}(\mathcal{I}_S, \mathcal{O}_C)$ is surjective, we therefore get a surjective tangent map of $pr_1$ (and in fact an injective map of obstruction spaces by [**37**]) from which we get the smoothness of $pr_1$ at $(C \subset S \subset \mathbb{P}^N)$. Now using the exact sequence of Lemma 7.1 (i), we see that the injectivity of $H^1(\mathcal{N}_S \otimes \mathcal{I}_{C|S}) \to H^1(\mathcal{N}_S)$ is equivalent to the surjectivity of $\operatorname{Hom}(\mathcal{I}_S, \mathcal{O}_S) \to \operatorname{Hom}(\mathcal{I}_S, \mathcal{O}_C)$. Hence the theorem of [**37**] proves (iii) if we combine with (ii). □

THEOREM 9.4. *Let $(C \subset S \subset \mathbb{P}^N) \in D(p,q)$, $C$ a Cartier divisor on $S$, and suppose $S$ is unobstructed and satisfies $H^2(R,B,B)^\sim = 0$. Let $t$ be an integer such that $H^1(\mathcal{O}_S(C)(t)) = 0$, and suppose the relative Picard scheme Pic is formally smooth at $(\mathcal{O}_S(C), S)$. If $C_t \in |C + tH|$ is an arbitrary effective Cartier divisor inducing an injective map $H^1(\mathcal{N}_S \otimes \mathcal{I}_{C|S}(-t)) \to H^1(\mathcal{N}_S)$, then $C_t$ is unobstructed, and we have*

$$\begin{aligned}\dim_{C_t} \operatorname{Hilb}^p(\mathbb{P}^N) \ &= \ h^0(\mathcal{N}_S) + h^1(\mathcal{O}_S) - \dim Im\ \alpha_L - h^0(\mathcal{N}_S \otimes \mathcal{I}_{C|S}(-t)) \\ &\quad + h^0(\mathcal{O}_S(C)(t)) - 1.\end{aligned}$$

PROOF. It is easy to see that Pic is formally smooth at $(L, S)$, where $L = \mathcal{O}_S(C)$, if and only if it is formally smooth at $(L(t), S)$. Moreover, since $H^1(\mathcal{O}_S(C)(t)) = 0$ and the map $H^1(\mathcal{N}_S \otimes \mathcal{I}_{C_t|S}) \to H^1(\mathcal{N}_S)$ is injective, the rational map $D(p_t, q) \to \operatorname{Pic}$ and the projection $pr_1 : D(p_t, q) \to \operatorname{Hilb}^{p_t}$ are (formally) smooth at $(C_t \subset S \subset \mathbb{P}^N)$, (cf. the proof of 9.2), and we get the unobstructedness of $C_t$. Finally the fibers of these smooth maps are easy to identify (as $H^0(\mathcal{O}_S(C)(t))/k$ and $H^0(\mathcal{N}_S \otimes \mathcal{I}_{C|S}(-t))$, cf. [**42**], p. 324 and [**38**], (1.6)), and since $\dim_{(L(t),S)} \operatorname{Pic} = \dim_{(L,S)} \operatorname{Pic}$, we get

$$\dim_{C_t} \operatorname{Hilb}^p = \dim_{(L,S)} \operatorname{Pic} - h^0(\mathcal{N}_S \otimes \mathcal{I}_{C|S}(-t)) + h^0(\mathcal{O}_S(C)(t)) - 1.$$

Now, since any deformation $S_{k[\epsilon]}$ of $S$ to the dual numbers $k[\epsilon]$ maps, via $\alpha_L : H^0(\mathcal{N}_S) \to H^2(\mathcal{O}_S)$, onto the obstruction of deforming $L$ to $S_{k[\epsilon]}$, and since, for any $S_{k[\epsilon]}$ in the kernel of $H^0(\mathcal{N}_S) \to H^2(\mathcal{O}_S)$, we can deform $L$ in each tangent direction given by $H^1(\mathcal{O}_S)$ (cf. [**27**]), we get

$$\dim_{(L,S)} \operatorname{Pic} = h^0(\mathcal{N}_S) - \dim Im\ \alpha_L + h^1(\mathcal{O}_S)$$

by the formal smoothness of Pic at $(L, S)$. One may also deduce this formula for $\dim_{(L,S)} \operatorname{Pic}$ from the exact sequence of [**42**], p. 324, describing the tangent space of Pic at $(L, S)$. We get easily the dimension formula of the theorem, and we are done. □

REMARK 9.5. If $S$ is of codimension 2 in $\mathbb{P}^N$ and satisfies $G_0$, the condition $H^2(R,B,B)^\sim = 0$ always holds by Proposition 6.17. For the higher codimension case this vanishing (or a locally licci assumption on $S$; cf. Proposition 6.17) could be difficult to verify. Therefore we remark that Proposition 9.2 and Theorem 9.4 are still true without requiring the vanishing of this local algebra cohomology, provided we replace the injectivity assumptions on $H^1(\mathcal{N}_S \otimes \mathcal{I}_{C|S}(-t)) \to H^1(\mathcal{N}_S)$ by a harder claim; the injectivity of $\operatorname{Ext}^1(\mathcal{I}_S, \mathcal{I}_{C|S}(-t)) \to \operatorname{Ext}^1(\mathcal{I}_S, \mathcal{O}_S)$ (we have to replace $\mathcal{N}_S \otimes \mathcal{O}_C$ by $\mathcal{H}om(\mathcal{I}_S, \mathcal{O}_C)$ in the definition of $\alpha_L$ and $\alpha_{C|S}$ as well). Indeed considering the proof of these results and noting that the smoothness of $pr_1 : D(p,q) \to \operatorname{Hilb}^p$ only requires the surjectivity of $\operatorname{Hom}(\mathcal{I}_S, \mathcal{O}_S) \to \operatorname{Hom}(\mathcal{I}_S, \mathcal{O}_C)$ and the unobstructedness of $S$ by [**37**], Theorem 1.3.4, we conclude as required.

In next example, we will give a huge family of ACM curves in $\mathbb{P}^4$ for which Theorem 9.4 applies.

EXAMPLE 9.6. Let $C \subset \mathbb{P}^4$ be a smooth ACM curve lying on a smooth Del Pezzo surface $S \subset \mathbb{P}^4$. Then, $C$ is unobstructed and

$$\dim_C \operatorname{Hilb}^p(\mathbb{P}^4) = 25 - 2h^0(\mathcal{I}_{C|S}(2)) + h^0(\mathcal{O}_S(C)) = \begin{cases} d+g+25 & \text{if } d > 8 \\ 5d+1-g & \text{if } d \leq 8. \end{cases}$$

In fact, since $S$ is a smooth complete intersection scheme in $\mathbb{P}^4$ of codimension 2, the algebra cohomology group $H^2(R,B,B)$ of $S$ vanishes (cf. Proposition 6.17) and $S$ is unobstructed. Since $H^2(\mathcal{O}_S) = 0$, the map $\alpha_L : H^0(\mathcal{N}_S) \to H^2(\mathcal{O}_S)$ is surjective. So, applying Proposition 9.2, we get that the relative Picard scheme Pic is formally smooth at $(\mathcal{O}_S(C), S)$. Moreover, $H^1(\mathcal{O}_S(C)) = H^1(\mathcal{N}_S \otimes \mathcal{I}_{C|S}) = 0$ (indeed,

$$\begin{aligned} H^1(\mathcal{O}_S(C)) &= H^1(\mathcal{O}_S(K-C)) \\ &= H^1(\mathcal{O}_S(-H-C)) \\ &= H^1(\mathcal{I}_{C|S}(-H)) \\ &= H^1(\mathcal{I}_C(-1)) \\ &= 0, \end{aligned}$$

and $H^1(\mathcal{N}_S \otimes \mathcal{I}_{C|S}) = 2H^1(\mathcal{I}_{C|S}(2H)) = H^1(\mathcal{I}_C(2)) = 0$).

So we are ready to apply Theorem 9.4 and we get the unobstructedness of $C$ and

$$\dim_C \operatorname{Hilb}^p(\mathbb{P}^4) = 25 - 2h^0(\mathcal{I}_{C|S}(2)) + h^0(\mathcal{O}_S(C))$$

because $h^0(\mathcal{N}_S) = 2h^0(\mathcal{O}_S(2H)) = 26$ (by the Riemann-Roch Theorem) and $h^1(\mathcal{O}_S) = \dim \operatorname{im} \alpha_L = 0$. Since, $H^0(\mathcal{I}_{C|S}(2H)) = H^0(\mathcal{O}_S(2H-C)) = 0$ if $d > 8$ and $H^0(\mathcal{O}_S(C)) = \chi(\mathcal{O}_S(C)) = d+g$; we easily conclude that

$$\dim_C \operatorname{Hilb}^p(\mathbb{P}^4) = \begin{cases} d+g+25 & \text{if } d > 8 \\ 5d+1-g & \text{if } d \leq 8. \end{cases}$$

As indicated earlier, the injectivity of $H^1(\mathcal{N}_S \otimes \mathcal{I}_{C|S}(-t)) \to H^1(\mathcal{N}_S)$ in Theorem 9.4 is related to the exactness of the complex (7.2). In fact, we have

PROPOSITION 9.7. *Let $(C \subset S \subset \mathbb{P}^N) \in D(p,q)$, $C$ a Cartier divisor on $S$, and suppose $C$ and $S$ are ACM schemes and $n = \dim C > 0$. As before, let $s(C|S)$ be the minimum degree of the generators of $I_{C|S}$ and let*

$$a = \max\{v \mid H^{n+1}(\mathcal{O}_S(v)) \neq 0\} + 1.$$

*If $t \geq n+4+a-s(C|S)$, then*

$$H^i(\mathcal{N}_S \otimes \mathcal{I}_{C|S}(-t)) = 0 \quad \text{and} \quad \operatorname{Ext}^i(\mathcal{I}_S, \mathcal{I}_{C|S}(-t)) = 0$$

*for $i = 0$ and 1. In particular, if $N - n = 3$ the complex (7.2) is exact for $v = -t$.*

PROOF. Suppose $N = n+3$. By (7.4) and (7.5), the complex (7.2) is exact for $v = -t$ if and only if $H^1(\mathcal{N}_S \otimes \mathcal{I}_{C|S}(-t)) = 0$. Indeed, if we can show that the following term of (7.2) vanishes:

$$\oplus_i H^0(\mathcal{I}_{C|S}(q_i - t)) = 0, \tag{9.2}$$

then $H^1(\mathcal{N}_S \otimes \mathcal{I}_{C|S}(-t))$ and $\operatorname{Ext}^1(\mathcal{I}_S, \mathcal{I}_{C|S}(-t))$ both vanish. To prove (9.2), it suffices to prove $q_i - t < s(C|S)$ for all $i$, i.e. $t > \max\{q_i\} - s(C|S)$ by the definition of $s(C|S)$. Since the numbers $q_i$ are given by (7.1) and since

$$0 \to H^{n+2}(\mathcal{I}_S(v)) \to \oplus_i H^{n+3}(\mathcal{O}_{\mathbb{P}^{n+3}}(v-q_i)) \to \oplus_i H^{n+3}(\mathcal{O}_{\mathbb{P}^{n+3}}(v-p_i))$$

is exact, we see that $H^{n+1}(\mathcal{O}_S(v)) \cong H^{n+2}(\mathcal{I}_S(v))$ vanishes if $v - q_i \geq -n-3$ for all $i$. Indeed, $a = \max\{q_i\} - n - 3$ because $\max\{q_i\} > \max\{p_i\}$, and hence we have (9.2) and the desired vanishing for any integer $t \geq n+4+a-s(C|S)$.

Finally, since for such $t$ we also have $H^0(\mathcal{I}_{C|S}(p_i - t)) = 0$ for all $i$, we see that $H^0(\mathcal{N}_S \otimes \mathcal{I}_{C|S}(-t)) = \operatorname{Ext}^0(\mathcal{I}_S, \mathcal{I}_{C|S}(-t)) = 0$ because these groups are in general a subgroup of $\oplus_i H^0(\mathcal{I}_{C|S}(p_i - t))$.

If $N < n+3$ the result easily follows. Assume $N > n+3$. So, a minimal resolution of $I_S$ is given by

$$0 \to \oplus_i R(-n_{r,i}) \to \ldots \to \oplus_i R(-n_{2,i}) \to \oplus_i R(-n_{1,i}) \to I_S \to 0 \tag{9.3}$$

where $r = N - n - 1$. Note that by minimality

$$\max\{n_{r,i}\} > \ldots > \max\{n_{2,i}\} > \max\{n_{1,i}\}. \tag{9.4}$$

Moreover splitting the sheafification of (9.3) into short exact sequences and taking cohomology, we see that

$$\begin{aligned} 0 \to H^{n+1}(\mathcal{O}_S(v)) \cong H^{n+2}(\mathcal{I}_S(v)) \to \textstyle\bigoplus_i H^N(\mathcal{O}_{\mathbb{P}^N}(v - n_{r,i})) \\ \to \textstyle\bigoplus_i H^N(\mathcal{O}_{\mathbb{P}^N}(v - n_{r-1,i})) \end{aligned}$$

is exact. Hence by the definition of $a$ and (9.4),

$$a = \max\{n_{r,i}\} - N = \max\{n_{r,i}\} - n - r - 1 \geq \max\{n_{2,i}\} - n - 3.$$

By applying $\operatorname{Hom}(-, \mathcal{I}_{C|S}(-t))$ to the sheafification of (9.3), we get the desired vanishing of $\operatorname{Ext}^i(\mathcal{I}_S, \mathcal{I}_{C|S}(-t))$. Moreover, since $\mathcal{I}_{C|S}$ is invertible, we have

$$\mathcal{H}om(\mathcal{I}_S, \mathcal{I}_{C|S}) \cong \mathcal{N}_S \otimes \mathcal{I}_{C|S},$$

and we get the vanishing of $H^i(\mathcal{N}_S \otimes \mathcal{I}_{C|S}(-t))$ for $i = 0$ and $1$ by the spectral sequence used in (7.5). □

REMARK 9.8. (i) By the final part of the proof above, we see that the vanishing of $\operatorname{Ext}^i(\mathcal{I}_S, \mathcal{I}_{C|S}(-t))$ for $i = 0, 1$ is true without assuming $C \subset S$ to be Cartier, keeping of course the other assumptions of Proposition 9.7.

(ii) In order to apply Theorem 9.4 in the non-ACM case it is interesting to prove a variation of Proposition 9.7 without ACM conditions. For this non-ACM case of subschemes in $\mathbb{P}^N$ we only remark that $H^1(\mathcal{O}_S(t))$ and $\operatorname{Ext}^i(\mathcal{I}_S, \mathcal{I}_{C|S}(-t))$, $i = 0,1$ vanish for $t \gg 0$. For the $\operatorname{Ext}^i$-groups, this follows again by applying $\operatorname{Hom}(-, \mathcal{I}_{C|S}(-t))$ to the first two terms of a projective resolution of $I_S$.

Using Proposition 9.7 and Remark 9.8 (ii), Theorem 9.4 and Remark 9.5 immediately imply

COROLLARY 9.9. *Let $(C \subset S \subset \mathbb{P}^N) \in D(p,q)$, $C$ a Cartier divisor on $S$ of dimension $n = \dim C > 0$, and suppose the map $\alpha_{\mathcal{O}_S(C)}$ is surjective and $S$ is unobstructed. Then any member $C_t$ of $|C + tH|$ is unobstructed for $t \gg 0$, and we have*

$$\dim_{C_t} \operatorname{Hilb}^p = h^0(\mathcal{N}_S) + h^1(\mathcal{O}_S) - h^2(\mathcal{O}_S) + h^0(\mathcal{O}_S(C)(t)) - 1.$$

*Moreover, if $C$ and $S$ are ACM subschemes in $\mathbb{P}^N$ and as before*

$$a = \max\{v | H^{n+1}(\mathcal{O}_S(v)) \neq 0\} + 1$$

*and $s(C|S)$ is the minimal degree of the generators of $I_{C|S}$, then the conclusion above holds for any $t \geq \max\{\beta, n+4+a-s(C|S)\}$ where*

$$\beta = \max\{v | H^1(\mathcal{O}_S(C)(v)) \neq 0\} + 1 \quad (\beta = -\infty \text{ if } H^1_*(\mathcal{O}_S(C)) = 0).$$

The following corollary allows us to easily treat the examples we have in mind.

COROLLARY 9.10. *(cf. [**39**], Corollary 2.3) Let $C \subset \mathbb{P}^N$ be an ACM curve of degree $d$ and (arithmetic) genus $g$, sitting on a smooth rational ACM surface $S \subset \mathbb{P}^N$ of degree $s$. Let $K$ be the canonical divisor of $S$. If $t$ is any integer such that $t \geq 5 - s(C|S)$ and $t > (C \cdot K)/d$, then any member $C_t$ of $|C + tH|$ is an unobstructed curve and*

$$\dim_{C_t} \operatorname{Hilb}^p = h^0(\mathcal{N}_S) + \chi(\mathcal{O}_S(C)(t)) - 1.$$

*In particular, $C$ is unobstructed if $C \cdot K < 0$ and $s(C|S) \geq 5$. Moreover, if $S$ is nondegenerate in $\mathbb{P}^N$, then*

$$h^0(\mathcal{N}_S) = \chi(\mathcal{N}_S) = N^2 + 2N + 10 - 2K^2$$

*and*

$$\chi(\mathcal{O}_S(C)(t)) = g - C \cdot K + t(d + N - s/2) + t^2 s/2.$$

PROOF. Note that a smooth rational ACM surface in $\mathbb{P}^N$ is unobstructed because one may easily prove that $H^2(\mathcal{O}_S) = 0$, $H^2(\theta_S) = 0$ ( $\theta_S$ is the tangent sheaf) and $H^1(\mathcal{O}_S(1)) = 0$ leads to $H^1(\mathcal{N}_S) = 0$ [**39**], Corollary 1.7. Therefore, if we can prove $H^1(\mathcal{O}_S(C)(t)) = 0$, we get the unobstructedness of $C_t$ from Theorem 9.4 by combining Proposition 9.2 and Proposition 9.7. Now to see $H^1(\mathcal{O}_S(C)(t)) = 0$, we consider the exact sequence

$$0 \to \mathcal{O}_S \to \mathcal{O}_S(C) \to \mathcal{N}_{C|S} \to 0,$$

where $\mathcal{N}_{C|S} \cong \omega_C \otimes \omega_S^{\otimes(-1)}$. Twisting by $t$ and taking cohomology, it suffices to prove $H^1(\mathcal{N}_{C|S}(t)) \cong H^0(\omega_S(-t) \otimes \mathcal{O}_C)^\vee = 0$. Since the degree of $\omega_S(-t) \otimes \mathcal{O}_C$ is negative by assumption, we conclude easily.

Now the dimension formula follows by combining Theorem 9.4 and Proposition 9.7. Moreover $H^2(\mathcal{N}_S) = 0$ and

$$\chi(\mathcal{N}_S) = (N+1)\chi(\mathcal{O}_S(1)) - 2K^2 + (N+10)\chi(\mathcal{O}_S)$$

by [**39**], Proposition 1.6. If $S$ is nondegenerate, we have $\chi(\mathcal{O}_S(1)) = N+1$, and we conclude by the Riemann-Roch Theorem. □

EXAMPLE 9.11. (Curves on a Castelnuovo surface S in $\mathbb{P}^4$; cf. Example 7.11)

a) Let $C$ be a curve of the system $(1; 0^8)$ of degree $d = C \cdot H = 4$ and arithmetic genus $g = 0$. Then $s(C|S) = 2$ and any curve $C_t$ of the linear system $(4t+1; 2t, t^7)$ is unobstructed and

$$\dim_{C_t} \operatorname{Hilb}^p = 34 + 11t/2 + 5t^2/2 = 5\deg(C_t) + 1 - g(C_t) + (5t^2 - 17t + 13)$$

for $t \geq 3$ by Corollary 9.10 (in which we leave to the reader to find $\deg(C_t)$ and $g(C_t)$ and check the equality to the right). Moreover, if $t \leq 1$, we have $H^1(\mathcal{N}_{C_t}) = 0$ for smooth curves $C_t$ because $2g(C_t) - 2 < \deg(C_t)$. We have used Macaulay [**5**] to check the remaining case $C_2$. We got $H^1(\mathcal{N}_{C_2}) = 0$ for a general enough $C_2$, so any general $C_t$ is unobstructed for $t \geq 0$. Finally we remark that Corollary 7.4 may be used to see the vanishing of ${}_0H^2(R, A_t, A_t)$ for $t \geq 4$, leading to another proof of the unobstructedness of $C_t$ in this case.

b) The other curves we considered in Example 7.11 can be treated similarly. Indeed let $C$ be a curve of the system $(1;1,0^7)$. Since $s(C|S)=1$, *any* curve $C_t$ of $(4t+1;2t+1,t^7)$ is unobstructed provided $t \geq 4$. Correspondingly, if $C$ is a section of $(0;0^7,-1)$ (resp. of $(4;3,1^7)$ or of any linear system containing an ACM curve), then any $C_t$ of the system $|C+tH|$ is unobstructed provided $t \geq 4$ by Corollary 9.10. In each case $\dim_{C_t} \text{Hilb}^p$ is easily computed.

REMARK 9.12. There exist ACM curves in $\mathbb{P}^4$ of degree 4 and genus 0 which are obstructed, cf. [**50**]. As an example we remark that the curve $C$ in $\mathbb{P}^4 = \text{Proj}(k[x_0,x_1,x_2,x_3,x_4])$ given by $I_C = (x_0,x_1,x_2)^2$ is an obstructed standard determinantal curve; note that $C$ is not generically a complete intersection. However, the curve $C'$ given by the union of four lines in $\mathbb{P}^4$ meeting at a point is an example of an *obstructed* curve of degree 4 and arithmetic genus 0 which is even generically smooth (cf. [**59**], Theorem 4.2). One can check that $h^0(\mathcal{N}_{C'})=24$, while the dimension of the corresponding Hilbert scheme is 21. Note that $C'$ cannot be locally licci either, by Proposition 6.17, because the vanishing of $H^2(R,R/I_{C'},R/I_{C'})^\sim$ and $H^1(\mathcal{N}_{C'})$ imply unobstructedness.

Moreover, there exist obstructed curves in the class of Buchsbaum curves with a 1-dimensional Hartshorne-Rao module (i.e. the "closest class to the ACM class"), [**39**], Example 2.4. Among the last mentioned examples, there even exist smooth curves sitting as a Cartier divisors on a smooth K3 surface in $\mathbb{P}^4$.

On the other hand Corollary 9.10 and the results above (together with [**39**], Remark 2.2 and Corollary 2.3 which at least takes care of curves on a K3 surface as well) point in a direction that ACM curves obtained by adding hyperplane sections "normally" are unobstructed. Since another main class considered in this paper (determinantal codimension 3 subschemes) also seems to define unobstructed curves, cf. Proposition 10.7, we are led to pose the following two interesting questions:

1. Is every local complete intersection (resp. smooth) ACM curve in $\mathbb{P}^4$ unobstructed?
2. Is every smooth ACM curve contained in a smooth ACM surface in $\mathbb{P}^4$ unobstructed?

Since our investigations so far still leave too many classes to study, we are not yet ready to state affirmative answers to these questions as conjectures. However, we especially expect Question 2 to have a positive answer under weak or no extra assumptions, as Corollary 9.10 partially indicates. We also have the following result in this direction.

PROPOSITION 9.13. *Let $C \subset \mathbb{P}^4$ be a smooth, ACM curve lying on a smooth, rational ACM surface $S \subset \mathbb{P}^4$. If $\deg(C) \leq 8$ or $\deg(C) \geq 24$ then $C$ is unobstructed.*

PROOF. By Example 9.6 any smooth ACM curve $C \subset \mathbb{P}^4$ on a Del Pezzo surface $S$ is unobstructed. So, we will assume that S is a cubic scroll or a Castelnuovo surface or a Bordiga surface. If $d \leq 8$ it easily follows from the description of ACM curves given in Chapter 8 that $2g(C)-2 \leq d$. Thus, either $H^1(\mathcal{O}_C(1))=0$ and hence $H^1(\mathcal{N}_C)=0$ or else $C$ is an arithmetically Gorenstein curve with $(d,g)=(8,5)$ on a Bordiga surface. Therefore, $C$ is unobstructed.

On the other hand, for any minimal ACM curve $C_0$ on $S$ we have $\deg(C_0+\mu H) > 23$ implies $\mu \geq 4$, unless $S$ is a Bordiga surface and $\deg C_0 \geq 6$, in which case $\mu \geq 3$ and $s(C_0|S) \geq 2$ (cf. the classification of minimal curves in

Chapter 8). Therefore, applying Corollary 9.10 (and Theorem 9.4 if $C \in |\mu H|$) we get that any curve $C$ on $S$ of degree $\geq 24$ is unobstructed. □

If $(C_t \subset S \subset \mathbb{P}^N)$ is a general point of an irreducible (non-embedded) component of $D(p_t, q)$ and the map $H^1(\mathcal{N}_S \otimes \mathcal{I}_{C|S}(-t)) \to H^1(\mathcal{N}_S)$ of Theorem 9.4 is not injective, then the dimension formula of the Theorem has a natural interpretation as the dimension of a certain strata of $\mathrm{Hilb}^p = \mathrm{Hilb}^p(\mathbb{P}^N)$, whose codimension is bounded above by the dimension of the kernel of $H^1(\mathcal{N}_S \otimes \mathcal{I}_{C|S}(-t)) \to H^1(\mathcal{N}_S)$. Indeed, looking at the proof of Theorem 9.4, there is in general no particular reason for the natural projection $pr_1 : D(p_t, q) \to \mathrm{Hilb}^{p_t}$ to be smooth or dominating near $(C_t \subset S \subset \mathbb{P}^N)$, i.e. a general member of a component of $\mathrm{Hilb}^{p_t}$ containing $(C_t \subset \mathbb{P}^N)$ might not sit on any deformation of $S \subset \mathbb{P}^N$. To understand this situation, let $t = 0$ and put

$$\delta(C \subset S) = \dim \ker \, [H^1(\mathcal{N}_S \otimes \mathcal{I}_{C|S}) \to H^1(\mathcal{N}_S)]$$

and $W(C \subset S) = pr_1(W)$, where $W$ is the irreducible component of $D(p, q)$ containing a general $(C \subset S \subset \mathbb{P}^N)$. Then $W(C \subset S)$ consists of members $(C' \subset \mathbb{P}^N)$ of $\mathrm{Hilb}^p$ for which there exists a member $(S' \subset \mathbb{P}^N)$ of $\mathrm{Hilb}^q$ such that $C' \subset S'$ is in $W$. We have

PROPOSITION 9.14. *Let $(C \subset S \subset \mathbb{P}^N)$ be a general point of an irreducible non-embedded component of $D(p, q)$, and suppose $C$ is a Cartier divisor on $S$ and $S$ is unobstructed and satisfies $H^2(R, B, B)^{\sim} = 0$. If $H^1(\mathcal{O}_S(C)) = 0$ and if* Pic *is formally smooth at $(\mathcal{O}_S(C), S)$, then*

$$\begin{aligned} \dim_C W(C \subset S) &= h^0(\mathcal{N}_S) + h^1(\mathcal{O}_S) - \dim Im \, \alpha_{\mathcal{O}_S(C)} - h^0(\mathcal{N}_S \otimes \mathcal{I}_{C|S}) \\ &\quad + h^0(\mathcal{O}_S(C)) - 1 \end{aligned}$$

*and*

$$\dim_C \mathrm{Hilb}^p(\mathbb{P}^N) - \dim_C W(C \subset S) \leq \delta(C \subset S).$$

*Moreover if* $\dim_C \mathrm{Hilb}^p(\mathbb{P}^N) = \dim_C W(C \subset S) + \delta(C \subset S)$, *then $C$ is unobstructed. The converse holds if in addition $\alpha_{\mathcal{O}_S(C)} : H^0(\mathcal{N}_S) \to H^2(\mathcal{O}_S)$ is surjective.*

PROOF. Proceeding as in the proof of Theorem 9.4, slightly modified, we see that $D(p, q)$ is smooth at $(C \subset S \subset \mathbb{P}^N)$ and

$$\dim_{C \subset S} W = h^0(\mathcal{N}_S) + h^1(\mathcal{O}_S) - \dim Im \, \alpha_{\mathcal{O}_S(C)} + h^0(\mathcal{O}_S(C)) - 1.$$

Endow $W(C \subset S)$ with its reduced scheme structure. Applying the Theorem of generic smoothness to the restriction $pr_1|_W : W \to W(C \subset S)$ of $pr_1 : D(p, q) \to \mathrm{Hilb}^p$, and recalling that the smooth fibers of $pr_1$ (and hence of $pr_1|_W$) have fiber dimension $h^0(\mathcal{N}_S \otimes \mathcal{I}_{C|S})$ (cf. the proof of Theorem 9.4), we get the first dimension formula.

Next we recall ([**38**], (1.6)) that the tangent map $T_{pr_1}$ of $pr_1$ is given as in (6.1), after having taken global sections. Hence we have $T_{pr_1} : H^0(A^1_{C|S}) \to H^0(\mathcal{N}_C)$ and since (6.1) is cartesian, we get $\mathrm{coker}\, T_{pr_1} \hookrightarrow \ker \, [H^1(\mathcal{N}_S \otimes \mathcal{I}_{C|S}) \to H^1(\mathcal{N}_S)]$. Combining with the smoothness of $D(p, q)$ at $(C \subset S \subset \mathbb{P}^N)$, which gives $\dim_{C \subset S} W = h^0(A^1_{C|S})$, and the proven formula for $\dim_C W(C \subset S)$, we get

$$h^0(\mathcal{N}_C) \leq \dim_C W(C \subset S) + \delta(C \subset S). \tag{9.5}$$

This proves the codimension formula because $\dim_C \mathrm{Hilb}^p \leq h^0(\mathcal{N}_C)$. Since unobstructedness is equivalent to $\dim_C \mathrm{Hilb}^p = h^0(\mathcal{N}_C)$, we also get the first part of the final statement. Moreover, by $H^1(\mathcal{O}_S(C)) = 0$, the surjectivity of $\alpha_{\mathcal{O}_S(C)}$ is equivalent to the surjectivity of $\alpha_{C|S}$, which by some diagram chasing or the exact sequence of [**38**] (1.11), imply the equality in (9.5), and we conclude easily. □

To apply Proposition 9.14, we remark that the assumption "$(C \subset S \subset \mathbb{P}^N)$ a general point of an irreducible non-embedded component of $D(p,q)$" is supposed to be fulfilled provided $(\mathcal{O}_S(C), S)$ is general enough in Pic. Indeed, we may think of this as a consequence of the continuity of the rational map $D(p,q) \to$ Pic at $(\mathcal{O}_S(C), S)$. We illustrate Proposition 9.14 by treating some exceptional cases of the preceding Example 9.11.

EXAMPLE 9.15. (Curves on a general Castelnuovo surface $S$ in $\mathbb{P}^4$). Let $C$ be a general curve of the system $(1;1,0^7)$ of degree $d = C \cdot H = 2$ and genus $g = 0$. Note that $C$ is a twisted canonical divisor because $K + H$ also corresponds to $(1;1,0^7)$. In Example 9.11 we saw that any curve $C_t$ of $(4t+1; 2t+1, t^7)$ was unobstructed provided $t \geq 4$. Now we consider curves $C_t$ with $0 \leq t \leq 3$. With notation as in Proposition 9.14, we claim that

| $t$ | $h^0(\mathcal{N}_S \otimes \mathcal{I}_{C_t\|S})$ | $h^1(\mathcal{N}_S \otimes \mathcal{I}_{C_t\|S})$ | $\dim_{C_t} \mathrm{Hilb}^{p_t} - \dim_{C_t} W(C_t \subset S)$ |
|---|---|---|---|
| 0 | 22 | 0 | 0 |
| 1 | 7 | 1 | 1 |
| 2 | 1 | $1 + \dim L^0(C_2)_0 \leq 3$ | $\leq 3$ |
| 3 | 0 | $-1 + \dim L^0(C_3)_0 \leq 4$ | $\leq 4$ |
| $\geq 4$ | 0 | 0 | 0 |

To compute $h^0(\mathcal{N}_S \otimes \mathcal{I}_{C_t|S}) = h^0(\mathcal{N}_S \otimes \mathcal{I}_{C|S}(-t))$, we remark that

$$\begin{aligned} \mathcal{N}_S \otimes \mathcal{I}_{C|S} &\cong \mathcal{E}xt^1(\mathcal{I}_S, \mathcal{I}_S) \otimes \mathcal{O}_S(-C) \\ &\cong \mathcal{E}xt^1(\mathcal{I}_S, \mathcal{O}_{\mathbb{P}}) \otimes \mathcal{I}_S \otimes \mathcal{O}_S(-K-H) \\ &\cong \mathcal{I}_S \otimes \mathcal{O}_S(4) \end{aligned}$$

because

$$\mathcal{E}xt^1(\mathcal{I}_S, \mathcal{O}_{\mathbb{P}}(-5)) \cong \mathcal{E}xt^2(\mathcal{O}_S, \mathcal{O}_{\mathbb{P}}(-5)) \cong \mathcal{O}_S(K).$$

Hence

$$H^0(\mathcal{N}_S \otimes \mathcal{I}_{C|S}(-t)) \cong H^0(\mathcal{I}_S/\mathcal{I}_S^2(4-t)) \cong (I_S/I_S^2)_{4-t},$$

where the last isomorphism relies on the fact that $\mathrm{pd}(I_S/I_S^2) = 3$, cf. [**3**], Theorem 3.2 or Remark 10.9. Since

$$0 \to R(-4)^{\oplus 2} \to R(-3)^{\oplus 2} \oplus R(-2) \to I_S \to 0,$$

we easily get the values of $h^0(\mathcal{N}_S \otimes I_{C_t|S})$ above.

Next we observe that the homology groups of the complex (7.2) (enlarged so that it starts and ends with 0's) are just $L^0(C)_v$ and $H^i(\mathcal{N}_S \otimes \mathcal{I}_{C|S}(v))$ for $i = 0,1$. Some easy computations show $h^0(\mathcal{I}_{C|S}(v)) = 2, 9, 21, 38$ for $v = 1,2,3,4$ respectively. Since $L^0(C_t)_0 \cong L^0(C)_{-t} = 0$ for $t = 0$ and 1 by Proposition 7.2, we get the numbers in the column of $h^1(\mathcal{N}_S \otimes \mathcal{I}_{C_t|S}) = h^1(\mathcal{N}_S \otimes \mathcal{I}_{C|S}(-t))$ above. Then the last column follows at once from Proposition 9.14 (in which the two cases of inequality, which can probably be improved, are left to the reader). Note that the column $h^0(\mathcal{N}_S \otimes \mathcal{I}_{C_t|S})$, Proposition 9.14 and Corollary 9.10 allow us to compute the precise dimension of the strata $W(C_t \subset S)$. Indeed we get

$$\dim W(C_t \subset S) = 33 - h^0(\mathcal{N}_S \otimes \mathcal{I}_{C_t|S}) + (7t + 5t^2)/2$$

which one may use to confirm the codimension column given by Proposition 9.14 after having sufficiently computed $\dim_{C_t} \mathrm{Hilb}^{p_t}$. In particular $W(C_1 \subset S)$ is a proper closed subset of $\mathrm{Hilb}^{p_1}$ and so is $W(C_2 \subset S)$ in $\mathrm{Hilb}^{p_2}$ provided $C_2$ is unobstructed.

A consequence of the codimensions we have computed is the following: A general curve in $\mathbb{P}^4$ of degree 2 and genus 0 sits on a Castelnuovo surface and belongs to the system $(1; 1, 0^7)$, while no general (resp. general unobstructed) curve in $\mathbb{P}^4$ of degree 7 (resp. 12) and genus 3 (resp. 11) sits on a Castelnuovo surface as a section of $(5; 3, 1^7)$ (resp. of $(9; 5, 2^7)$).

Working with the Hilbert scheme of points, the smoothness of $\mathrm{GradAlg}^H$ and its dimension at $(C \subset \mathbb{P}^N)$ are given by the following result, in which $B = R/I_S$, $A_t = R/I_{C_t}$ and $H_t = H_{C_t}$, and we have denoted by $\mathrm{GradAlg}(H_C, H_S)$ the subscheme of $D(p,q)$ obtained by deforming "pairs" $(C, S)$, $C \subset S \subset \mathbb{P}^N$, with constant Hilbert functions $H_C$ and $H_S$ respectively.

THEOREM 9.16. *Let $(C \subset S \subset \mathbb{P}^N) \in \mathrm{GradAlg}(H_C, H_S)$, $C$ a Cartier divisor on $S$ and suppose that $S$ is an $H_S$-unobstructed curve (so $C$ is a zeroscheme). Let $C_t \in |C + tH|$ be an arbitrary effective Cartier divisor such that*

$$ {}_0\mathrm{Hom}_R(I_{C_t|S}, H^2_{\mathfrak{m}}(I_{C_t})) = 0. $$

*If ${}_0\mathrm{Ext}^1_R(I_S, I_{C_t|S}) \to {}_0\mathrm{Ext}^1_R(I_S, B)$ is injective, then $C_t$ is $H_t$-unobstructed. If moreover $S$ is an ACM curve and a local complete intersection along $C_t$, then $C_t$ is unobstructed and*

$$ \begin{aligned} \dim_{C_t} \mathrm{GradAlg}^{H_t} &= h^0(\mathcal{N}_{C_t|S}) + h^0(\mathcal{N}_S) - h^0(\mathcal{N}_S \otimes \mathcal{I}_{C|S}(-t)), \\ \dim_{C_t} \mathrm{Hilb}^{p_t} - \dim_{C_t} \mathrm{GradAlg}^{H_t} &= h^1(\mathcal{N}_S \otimes \mathcal{I}_{C|S}(-t)) - h^1(\mathcal{N}_S) \end{aligned} $$

*where $h^0(\mathcal{N}_{C_t|S}) = \chi(\mathcal{O}_S(C)(t)) - \chi(\mathcal{O}_S)$.*

PROOF. We claim that $\mathrm{GradAlg}(H_C, H_S)$ is smooth at $(C \subset S \subset \mathbb{P}^N)$ and ${}_0H^2(B, A, A) = 0$ provided ${}_0\mathrm{Hom}_R(I_{C|S}, H^2_{\mathfrak{m}}(I)) = 0$. Indeed by [**40**] Lemma 3.5, there is an exact sequence

(9.6)

$$ 0 \to {}_0H^1(B, A, A) \to H^0(\mathcal{N}_{C|S}) \to {}_0\mathrm{Hom}_R(I_{C|S}, H^2_{\mathfrak{m}}(I)) \to {}_0H^2(B, A, A) \to 0 $$

because $H^1(\mathcal{N}_{C|S}) = 0$ and $C$ is a Cartier divisor on $S$. The assumption that ${}_0\mathrm{Hom}_R(I_{C|S}, H^2_{\mathfrak{m}}(I)) = 0$ implies therefore ${}_0H^2(B, A, A) = 0$. If $T' \to T$ is a surjection of local Artin $k$-algebras with residue fields $k$, and $R \otimes T \to B_T \to A_T$ is any (flat) deformation of $R \to B \to A$ to $T$, we can deform $R \otimes T \to B_T$ further to $T'$ by the $H_S$-unobstructedness of $S$, and $B_T \to A_T$ further to $B_{T'}$ by ${}_0H^2(B, A, A) = 0$, and the claim is proved.

Next we claim that the projection $\mathrm{GradAlg}(H_C, H_S) \to \mathrm{GradAlg}^{H_C}$, essentially given by

$$ (C \subset S \subset \mathbb{P}^N) \mapsto (C \subset \mathbb{P}^N), $$

is smooth at $(C \subset S \subset \mathbb{P}^N)$ provided ${}_0\mathrm{Ext}^1_R(I_S, I_{C|S}) \to {}_0\mathrm{Ext}^1_R(I_S, B)$ is injective. Indeed the tangent map $T_{pr_1}$ of $\mathrm{GradAlg}(H_C, H_S) \to \mathrm{GradAlg}^{H_C}$ is given by a cartesian diagram as in (6.1) where we replace any sheaf $\mathcal{H}om(-,-)$ with the corresponding graded piece ${}_0\mathrm{Hom}(-,-)$ of the graded $\mathrm{Hom}(-,-)$. Hence we see that $T_{pr_1}$ is surjective if ${}_0\mathrm{Hom}_R(I_S, B) \to {}_0\mathrm{Hom}_R(I_S, A)$ is surjective. Since we can easily deduce this surjectivity from the injectivity of ${}_0\mathrm{Ext}^1_R(I_S, I_{C|S}) \to {}_0\mathrm{Ext}^1_R(I_S, B)$, the claim follows from [**37**], 1.4. (The fact that this follows was explained in the proof of Proposition 9.2 (iii) in a similar case. However, since we already know that $D(p,q)$ is smooth at $(C \subset S \subset \mathbb{P}^N)$, the conclusion follows directly from the surjectivity of $T_{pr_1}$ by [**26**], (17.11.1).) We get the $H_t$-unobstructedness of $C_t$ as well by combining the proven claims.

To get the first dimension formula, we observe that

$$\dim_{C_t} \operatorname{GradAlg}^{H_t} = \dim {}_0H^1(R, A_t, A_t)$$

by $H_t$-unobstructedness. Moreover, there is an exact sequence

$$(9.7) \qquad 0 \to {}_0H^1(B, A_t, A_t) \to {}_0H^1(R, A_t, A_t) \to {}_0H^1(R, B, A_t) \to 0$$

because ${}_0H^2(B, A_t, A_t) = 0$. Combining with (9.6) and recalling ${}_0H^1(R, B, A_t) \cong {}_0\operatorname{Hom}_R(I_S, A_t)$, we get $\dim_{C_t} \operatorname{GradAlg}^{H_t} = h^0(\mathcal{N}_{C_t|S}) + \dim {}_0\operatorname{Hom}_R(I_S, A_t)$. Now, for an ACM curve $S$, we have $H^0(\mathcal{N}_S) \cong {}_0\operatorname{Hom}_R(I_S, B)$ and hence

$$H^0(\mathcal{N}_S \otimes \mathcal{I}_{C_t|S}) \cong {}_0\operatorname{Hom}_R(I_S, I_{C_t|S}),$$

and we conclude by the injectivity assumption on the ${}_0\operatorname{Ext}^1$-groups.

Finally the sheafified version of (9.7) on the $H^2$-level shows $H^2(R, A_t, A_t)^{\sim} = 0$ because $S \subset \mathbb{P}^N$ is a local complete intersection along $C_t$, i.e. $C_t$ is unobstructed because $H^1(\mathcal{N}_{C_t}) = 0$. By taking global sections of the sheafified version of (9.7), it follows that $\dim_{C_t} \operatorname{Hilb}^{p_t} = h^0(\mathcal{N}_{C_t}) = h^0(\mathcal{N}_{C_t|S}) + h^0(\mathcal{N}_S \otimes \mathcal{O}_{C_t})$. Combining with the proven formula for $\dim_{C_t} \operatorname{GradAlg}^{H_t}$ we get the theorem by taking cohomology of

$$0 \to \mathcal{N}_S \otimes \mathcal{I}_{C_t|S} \to \mathcal{N}_S \to \mathcal{N}_S \otimes \mathcal{O}_{C_t} \to 0,$$

recalling $H^1(\mathcal{N}_S \otimes \mathcal{O}_{C_t}) = 0$. □

COROLLARY 9.17. *Let $C \subset \mathbb{P}^N$ be a zero-dimensional Cartier divisor sitting on an unobstructed ACM curve $S \subset \mathbb{P}^N$, and suppose $s(C|S) \geq \max\{a, e\}$ where $e = \max\{v | H^1(\mathcal{I}_C(v)) \neq 0\} + 1$ and $a = \max\{v | H^1(\mathcal{O}_S(v)) \neq 0\} + 1$. If $t$ is any integer such that $t \geq 4 + a - s(C|S)$, then any member $C_t$ of $|C + tH|$ is $H_t$-unobstructed. Moreover, if $S$ is a local complete intersection along $C_t$, then $C_t$ is unobstructed as well. In particular if $s(C|S) \geq \max\{a, e\}$ and $s(C|S) \geq 4 + a$, then $C$ is unobstructed in both senses.*

PROOF. Arguing exactly as in Proposition 9.7, we see that ${}_0\operatorname{Ext}^i_R(I_S, I_{C_t|S}) = 0$ for $i = 0$ and $1$ if $t \geq 4 + a - s(C|S)$ (valid also in codimension larger than 3). Combining this with Remark 7.9, cf. the proof of Corollary 7.10, we deduce ${}_0\operatorname{Hom}_R(I_{C_t|S}, H^2_{\mathfrak{m}}(I_{C_t})) = 0$ from $s(C|S) \geq \max\{a, e\}$. Then we get the corollary from Theorem 9.16. □

EXAMPLE 9.18. In Example 9.11 we considered curves $C$ (of degree 1, 2, 3 and 4) on a Castelnuovo surface in $\mathbb{P}^4$ (of degree 5, sectional genus 2 and minimal resolution $0 \to R(-4)^{\oplus 2} \to R(-3)^{\oplus 2} \oplus R(-2) \to I_S \to 0$), on which we added appropriate hyperplane sections to $C$. Similarly, diminishing the dimension by 1 everywhere, one may consider sets $C$ of $n$ reduced points (with $n = 1, 2, 3$ and 4) on a smooth ACM curve $S$ of degree 5 and genus 2 in $\mathbb{P}^3$ whose homogeneous ideal $I_S$ in $R = k[x_0, x_1, x_2, x_3]$ is given by

$$0 \to R(-4)^{\oplus 2} \to R(-3)^{\oplus 2} \oplus R(-2) \to I_S \to 0.$$

a) If $n = 4$ and $C$ has Hilbert function $H = (h_0, h_1, h_2, ...) = (1, 4, 4, 4..)$, where $h_v = \dim A_v$, one easily computes any number of Corollary 9.17. Indeed we get $a = 1$, $e = 1$, $s(C|S) = 2$, and so $C_t$ is $H_t$-unobstructed as well as unobstructed for any $t \geq 3$.

b) If $n = 2$ and $H = (1, 2, 2..)$, we get $a = 1$, $e = 1$, $s(C|S) = 1$, i.e. $C_t$ is unobstructed in both senses for any $t \geq 4$ by Corollary 9.17. We conclude in the same way for $n = 1$, $H = (1, 1, 1, ..)$, and for $n = 3$, $H = (1, 3, 3, ..)$.

We want to point out that Theorem 9.4 and 9.16 (with $t = 0$) contain a solution of the interesting problem of deciding when every member in a linear system is unobstructed provided one member is. Indeed the theorems and Proposition 9.2 imply at once:

COROLLARY 9.19. *Let $(C \subset S \subset \mathbb{P}^N) \in \mathrm{GradAlg}(H_C, H_S)$, $C$ a Cartier divisor on $S$ and suppose $S$ is unobstructed (resp. is an $H_S$-unobstructed curve) and satisfies $H^2(R, B, B)^\sim = 0$. Moreover suppose $H^1(\mathcal{N}_S \otimes \mathcal{I}_{C|S}) \to H^1(\mathcal{N}_S)$ (resp. ${}_0\mathrm{Ext}^1_R(I_S, I_{C|S}) \to {}_0\mathrm{Ext}^1_R(I_S, B)$) is injective and $H^1(\mathcal{O}_S(C)) = 0$ (resp. ${}_0\mathrm{Hom}_R(I_{C|S}, H^2_{\mathfrak{m}}(I)) = 0$). Then any member $C'$ of $|C|$ is unobstructed (resp. $H_C$-unobstructed) provided $C$ is unobstructed (resp. $H_C$-unobstructed).*

Now we will relate $h^0(\mathcal{N}_X)$ to $h^0(\mathcal{N}_{X'})$, where $X$ and $X'$ are two ACM schemes of dimension $n > 0$ algebraically linked by some complete intersection $Y \subset \mathbb{P}^N$ of type $f_1, \ldots, f_c$. The result completes Theorem 6.1, where we have seen that $H^i(\mathcal{N}_X) \cong H^i(\mathcal{N}_{X'})$ for $i = 1, \ldots, n-1$.

Since for unobstructed schemes, $h^0(\mathcal{N}_X)$ coincides with the dimension $\dim_X \mathrm{Hilb}^{p(s)}(\mathbb{P}^N)$ and, moreover, any scheme CI-linked to a unobstructed ACM scheme is unobstructed, this result gives us an effective method to compute $\dim_X \mathrm{Hilb}^{p(s)}(\mathbb{P}^N)$ for any licci scheme $X \subset \mathbb{P}^N$ (see Example 9.21).

PROPOSITION 9.20. *Let $X, X' \subset \mathbb{P}^N$ be ACM schemes of dimension $n > 0$ algebraically linked by some complete intersection $Y \subset \mathbb{P}^N$ of type $f_1, \ldots, f_c$. Then we have*

$$h^0(\mathcal{N}_X) = h^0(\mathcal{N}_{X'}) + \sum_i h^0(\mathcal{I}_{X'}(f_i)) - \sum_i h^0(\mathcal{I}_X(f_i))$$

*or, equivalently,*

$$h^0(\mathcal{N}_X) = h^0(\mathcal{N}_{X'}) + \sum_i h^0(\mathcal{O}_X(f_i)) - \sum_i h^0(\mathcal{O}_{X'}(f_i)).$$

PROOF. We consider the exact sequences (see (6.18))

$$0 \to \bigoplus_i \mathcal{I}_{X|Y}(f_i) \to A^1_{X|Y} \to \mathcal{N}_X \to 0 \tag{9.8}$$

and

$$0 \to \mathcal{I}_Y \to \mathcal{I}_X \to \mathcal{I}_{X|Y} \to 0. \tag{9.9}$$

Note that the global section functor $H^0$ is exact on (9.8) and (9.9); for (9.8) in the case $n = 1$, use (6.20). Hence we get

$$\begin{aligned} h^0(A^1_{X|Y}) &= h^0(\mathcal{N}_X) + \sum_i h^0(\mathcal{I}_{X|Y}(f_i)) \\ &= h^0(\mathcal{N}_X) + \sum_i h^0(\mathcal{I}_X(f_i)) - \sum_i h^0(\mathcal{I}_Y(f_i)) \end{aligned}$$

and

$$\begin{aligned} h^0(A^1_{X'|Y}) &= h^0(\mathcal{N}_{X'}) + \sum_i h^0(\mathcal{I}_{X'|Y}(f_i)) \\ &= h^0(\mathcal{N}_{X'}) + \sum_i h^0(\mathcal{I}_{X'}(f_i)) - \sum_i h^0(\mathcal{I}_Y(f_i)). \end{aligned}$$

Finally, from the liaison invariance of $h^0(A^1_{X|Y})$ we deduce

$$h^0(\mathcal{N}_X) = h^0(\mathcal{N}_{X'}) + \sum_i h^0(\mathcal{I}_{X'}(f_i)) - \sum_i h^0(\mathcal{I}_X(f_i)).$$

□

EXAMPLE 9.21. We know that there exist smooth ACM curves $C$ of degree $d = 5$ and genus $g = 1$ in $\mathbb{P}^4$ with a minimal free resolution

$$0 \to R(-5) \to R(-3)^{\oplus 5} \to R(-2)^{\oplus 5} \to I_C \to 0$$

(cf. the proof of Corollary 8.10). Since $H^1(\mathcal{O}_C(1)) = 0$, we have $h^1(\mathcal{N}_C) = 0$ and hence $C$ is unobstructed and $\dim_C \operatorname{Hilb}^{5s} = h^0(\mathcal{N}_C) = 5d + 1 - g = 25$ by Proposition 6.13.

Let $Y$ be a complete intersection of type $(3, 3, 6)$ containing $C$, and let $C'$ be the linked curve (which we may take to be a smooth ACM curve of degree $d'$ and genus $g'$). Using the mapping cone construction to the morphism induced by $I_Y \hookrightarrow I_C$ on their resolutions ([**62**], Proposition 2.5), we find the minimal free resolution of $I_{C'}$ to be

$$0 \to R(-10)^{\oplus 5} \to R(-9)^{\oplus 7} \oplus R(-6) \to R(-7) \oplus R(-6) \oplus R(-3)^{\oplus 2} \to I_{C'} \to 0.$$

By Proposition 2.5, $d' = \deg C' = 49$, $g' = 155$ and from the resolution we see easily that $h^0(\mathcal{O}_{C'}(3)) = 33$ and $h^1(\mathcal{O}_{C'}(6)) = 0$. Hence by Proposition 9.20, we get

$$\begin{aligned} \dim_{C'} \operatorname{Hilb}^{49s-154} &= h^0(\mathcal{N}_{C'}) \\ &= h^0(\mathcal{N}_C) - 2h^0(\mathcal{O}_C(3)) - h^0(\mathcal{O}_C(6)) + 2h^0(\mathcal{O}_{C'}(3)) \\ &\quad + h^0(\mathcal{O}_{C'}(6)) \\ &= 25 - 2\cdot 15 - 30 + 2 \cdot 33 + (49 \cdot 6 - 154) \\ &= 171. \end{aligned}$$

Since unobstructedness is preserved under CI-linkage of ACM schemes, $C'$ is unobstructed.

In principle Proposition 9.20 computes $\dim_C \operatorname{Hilb}^p$ for any licci subscheme $C \subset \mathbb{P}^N$, but direct formulas have been difficult to find. If $C \subset \mathbb{P}^{n+3}$ is a codimension 3, arithmetically Gorenstein subscheme, then $C$ is licci and a formula is known ([**44**], Theorem 2.6). Here we finish by pointing out a formula for $\dim_C \operatorname{Hilb}^p$ in two special cases where $C$ is licci (or weaker, where we suppose ${}_vH^0_{\mathfrak{m}}(K_A \otimes_R I_C) = 0$ for $v = 0$ and $v = n + 4$).

In the first case we combine Theorem 9.4 for $t = 0$ with Remark 9.3, where we see the connection between the surjectivity of $\alpha_L$ and the vanishing of ${}_0H^0_{\mathfrak{m}}(K_A(n+4) \otimes_R I_C)$. Leaving a few details to the reader, we easily prove the following result.

PROPOSITION 9.22. *Let $C \subset S$, $C$ a Cartier divisor on $S$, be two ACM subschemes in $\mathbb{P}^{n+3}$ of dimension $\dim C = n > 0$ and suppose $S$ satisfies $G_0$. If*

(i) $H^1(\mathcal{N}_S \otimes \mathcal{I}_{C|S}) \to H^1(\mathcal{N}_S)$ *is injective (true if* $s(C|S) \geq a+n+4$, *cf. Proposition 9.7),*

(ii) ${}_0\operatorname{Hom}(I_{C|S}, H_{\mathfrak{m}}^{n+1}(A)) = 0$ *(true if* $s(C|S) \geq e$ *by Remark 7.9) provided* $n = 1$, *and* $H^2(\mathcal{N}_{C|S}) = 0$ *provided* $n \geq 2$,

(iii) $\dim {}_0H_{\mathfrak{m}}^0(K_A(n+4) \otimes_R I_C) \leq \dim L^0(C)_0$, *and*

(iv) $H^1(\mathcal{O}_S(C)) = 0$

*then* $C$ *is unobstructed and*

$$\dim_C \operatorname{Hilb}^p = h^0(\mathcal{N}_S) + \chi(\mathcal{N}_{C|S}) = h^0(\mathcal{N}_S) + g - 1 - C \cdot K.$$

Curves which satisfy the conditions of Proposition 9.22 might have a minimal free resolution

$$0 \to \bigoplus_i R(-n_{3,i}) \to \bigoplus_i R(-n_{2,i}) \to \bigoplus_i R(-n_{1,i}) \to I_C \to 0 \tag{9.10}$$

which is far from being linear. In the other case, (9.10) is somewhat closer to being linear. Indeed, we have

PROPOSITION 9.23. *Let* $C$ *be an ACM, local complete intersection curve of degree* $d$ *and arithmetic genus* $g$ *in* $\mathbb{P}^4$, *let* $A = R/I_C$ *and suppose* ${}_vH_{\mathfrak{m}}^0(K_A \otimes_R I_C) = 0$ *for* $v = 0$ *and* $v = 5$. *If either*

(i) $\max n_{2,i} - \min n_{1,i} < 5$, *or*

(ii) $\max n_{3,i} - \min n_{2,i} < 5$

*then* $C$ *is unobstructed and*

$$\dim_C \operatorname{Hilb}^p = 5d + 1 - g + h^1(\mathcal{N}_C)$$

*where*

$$h^1(\mathcal{N}_C) = \sum h^1(\mathcal{O}_C(n_{1,i})) - \sum h^1(\mathcal{O}_C(n_{2,i})) + \sum h^1(\mathcal{O}_C(n_{3,i})).$$

*Moreover, in the case (i) (resp. (ii)), we also have*

$$h^1(\mathcal{N}_C) = \sum h^0(\mathcal{I}_C(n_{3,i} - 5)) \quad (\textit{resp. } h^1(\mathcal{N}_C) = \textstyle\sum h^1(\mathcal{O}_C(n_{1,i}))\ )$$

PROOF. Let $I = I_C$. Since $C$ is $H$-unobstructed by Remark 6.16, i.e. by ${}_0H^2(R, A, A) \cong {}_5H_{\mathfrak{m}}^0(K_A \otimes_R I) = 0$, and since $H$-unobstructedness and unobstructedness are equivalent for ACM curves, we get the unobstructedness of $C$ and hence

$$\dim_C \operatorname{Hilb}^p = h^0(\mathcal{N}_C) = 5d + 1 - g + h^1(\mathcal{N}_C)$$

by Proposition 6.13.

Now if we apply ${}_0\operatorname{Hom}(-, H_{\mathfrak{m}}^2(A))$ to (9.10), we get a complex

$$0 \to \bigoplus_i H^1(\mathcal{O}_C(n_{1,i})) \to \bigoplus_i H^1(\mathcal{O}_C(n_{2,i})) \to \bigoplus_i H^1(\mathcal{O}_C(n_{3,i})) \to 0$$

whose cohomology is ${}_0\operatorname{Ext}_R^i(I, H_{\mathfrak{m}}^2(A))$ for $i = 0,1,2$. Moreover, since

$${}_0H_{\mathfrak{m}}^1(K_A(5) \otimes_R I) \cong {}_0H_{\mathfrak{m}}^0(K_A \otimes_R I)^\vee$$

by the duality of Proposition 6.8, we get

$$H^1(\mathcal{N}_C) \cong {}_0\operatorname{Hom}_R(I, H_{\mathfrak{m}}^2(A))$$

by that proposition and by assumption.

If (i) holds, we claim that ${}_0\operatorname{Ext}^i_R(I, H^2_{\mathfrak{m}}(A)) = 0$ for $i = 1$ and 2. Indeed, using $H^2_{\mathfrak{m}}(A) \cong H^3_{\mathfrak{m}}(I)$, (6.23) and (6.26), we get

$$ {}_0\operatorname{Ext}^i_R(I, H^2_{\mathfrak{m}}(A)) \cong {}_0\operatorname{Ext}^{i+3}_{\mathfrak{m}}(I, I) \cong {}_{-5}\operatorname{Ext}^{2-i}_R(I, I)^\vee. $$

We have ${}_{-5}\operatorname{Hom}_R(I, I) \cong R_{-5} = 0$, and applying ${}_{-5}\operatorname{Hom}_R(-, I)$ to the resolution (9.10), we get ${}_{-5}\operatorname{Ext}^1_R(I, I) = 0$ because $(I_C)_{n_{2,i}-5} = 0$ by (i), i.e. the claim is proved. Note that the above argument also shows

$$ {}_{-5}\operatorname{Ext}^2_R(I, I) \cong \bigoplus_i (I_C)_{n_{3,i}-5}. $$

Now if we use the proven claim, and $H^1(\mathcal{N}_C) \cong {}_0\operatorname{Hom}(I, H^2_{\mathfrak{m}}(A))$, we easily get the first dimension formula for $h^1(\mathcal{N}_C)$.

To get the second dimension formula, it suffices to show that $H^1(\mathcal{N}_C) \cong {}_{-5}\operatorname{Ext}^2_R(I, I)^\vee$, i.e. ${}_0\operatorname{Hom}(I, H^2_{\mathfrak{m}}(A)) \cong {}_{-5}\operatorname{Ext}^2_R(I, I)^\vee$. Since $H^2_{\mathfrak{m}}(A) \cong H^3_{\mathfrak{m}}(I)$, the Hom-group is just ${}_0\operatorname{Ext}^3_{\mathfrak{m}}(I, I)$ by (6.23), and we conclude by the duality (6.26).

It remains to prove the Proposition when (ii) holds. Splitting (9.10) into two short exact sequences and taking cohomology, we see that $H^1(\mathcal{O}_C(n_{2,i})) = 0$ and $H^1(\mathcal{O}_C(n_{3,i})) = 0$. We get at once

$$ h^1(\mathcal{N}_C) = {}_0 hom_R(I, H^2_{\mathfrak{m}}(A)) = \sum h^1(\mathcal{O}_C(n_{1,i})) $$

and we conclude easily. □

EXAMPLE 9.24. The licci curve $C'$ of Example 9.21 satisfies (ii) of Proposition 9.23. Since

$$ h^1(\mathcal{O}_{C'}(3)) = h^0(\mathcal{O}_{C'}(3)) - (3d' + 1 - g') = 40, $$

we get

$$ h^1(\mathcal{N}_{C'}) = 2 \cdot h^1(\mathcal{O}_{C'}(3)) + h^1(\mathcal{O}_{C'}(6)) + h^1(\mathcal{O}_{C'}(7)) = 80 $$

and $\dim_{C'} \operatorname{Hilb}^{49t-154} = 5d' + 1 - g' + h^1(\mathcal{N}_{C'}) = 171$ by Proposition 9.23.

REMARK 9.25. (i) The vanishing of ${}_vH^0_{\mathfrak{m}}(K_A \otimes_R I_C)$ of Proposition 9.23 holds for licci curves. More generally, if an ACM, local complete intersection curve $C$ is in the CI-liaison class of a curve $C'$ satisfying $H^1(\mathcal{O}_{C'}(1)) = 0$ and $H^1(\mathcal{N}_{C'}) = 0$ (note that $H^1(\mathcal{O}_{C'}(1)) = 0$ implies $H^1(\mathcal{N}_{C'}) = 0$ if $C'$ is reduced), we get

$$ {}_5H^0_{\mathfrak{m}}(K_{A'} \otimes_R I_{C'}) = {}_5H^1_{\mathfrak{m}}(K_{A'} \otimes_R I_{C'}) = 0 $$

by the exact sequence of Proposition 6.8, and hence ${}_vH^0_{\mathfrak{m}}(K_A \otimes I_C) = 0$ for $v = 0$ and 5 by that Proposition. So Proposition 9.23 applies to such a curve if (i) or (ii) hold.

(ii) If we drop the assumptions (i) and (ii) of Proposition 9.23, a modification of its proof will still show

$$ \begin{aligned} h^1(\mathcal{N}_C) - h^0(\mathcal{N}_C(-5)) &= \sum h^1(\mathcal{O}_C(n_{1,i})) - \sum h^1(\mathcal{O}_C(n_{2,i})) + \sum h^1(\mathcal{O}_C(n_{3,i})) \\ &= \sum h^0(\mathcal{I}_C(n_{1,i} - 5)) - \sum h^0(\mathcal{I}_C(n_{2,i} - 5)) \\ &\quad + \sum h^0(\mathcal{I}_C(n_{3,i} - 5)). \end{aligned} $$

CHAPTER 10

# Dimension of families of determinantal subschemes

Our first goal in this chapter is to write down an upper bound for the dimension of the locus $W$ of good determinantal subschemes $C \subset \mathbb{P}^{n+c}$ (cf. Definition 3.1) of codimension $c$ inside the Hilbert scheme $\operatorname{Hilb}^{p(s)}(\mathbb{P}^{n+c})$, where $p(s) \in \mathbb{Q}[s]$ is the Hilbert polynomial of $C$. We can use the results of Chapter 9 in order to consider sharpness, and to see when $W$ is an irreducible component of $\operatorname{Hilb}^{p(s)}(\mathbb{P}^{n+c})$. One of the main results (Theorem 10.13) of this chapter is the unobstructedness of $C$ under certain conditions, and a dimension formula for $\operatorname{Hilb}^{p(s)}(\mathbb{P}^{n+c})$ at $(C \subset \mathbb{P}^{n+c})$. These assumptions are almost always satisfied if $c = 3$ and $n \geq 2$ (Proposition 10.7 and Corollary 10.15) and we can therefore find $\dim_C \operatorname{Hilb}^{p(s)}(\mathbb{P}^{n+c})$. In particular we prove that the mentioned bound is sharp (also when $c = 3$ and $n = 1$). If $c \geq 4$ the dimension formula of Theorem 10.13 is effective to use in some cases (Example 10.16).

Let $C \subset \mathbb{P}^{n+c}$ be a good determinantal scheme of codimension $c \geq 3$ (see **[22]** for the codimension 2 case) defined by the vanishing of the maximal minors of a $t \times (t+c-1)$ matrix $A = (f_{ij})_{i=1,\dots t}^{j=0,\dots,t+c-2}$ where $f_{ij} \in k[x_0, \dots, x_{n+c}]$ are homogeneous polynomials of degree $a_j - b_i$. The matrix $A$ defines a morphism of locally free sheaves

$$\varphi : \mathcal{F} := \bigoplus_{i=1}^{t} \mathcal{O}_{\mathbb{P}^{n+c}}(b_i) \longrightarrow \mathcal{G} := \bigoplus_{j=0}^{t+c-2} \mathcal{O}_{\mathbb{P}^{n+c}}(a_j)$$

with $b_1 \geq \dots \geq b_t$ and $a_0 \geq a_1 \geq \dots \geq a_{t+c-2}$. We may assume without loss of generality that $\varphi$ is minimal; i.e., $f_{ij} = 0$ for all $i, j$ with $b_i = a_j$. It is well known that the Eagon-Northcott complex associated to $\varphi$ gives us a minimal free resolution of $\mathcal{O}_C$. Up to shift, we have

$$\dots \longrightarrow \wedge^t \mathcal{G}^* \otimes \wedge^t \mathcal{F} \longrightarrow \mathcal{O}_{\mathbb{P}^{n+c}} \longrightarrow \mathcal{O}_C \longrightarrow 0.$$

Let $p(s) \in \mathbb{Q}[s]$ be the Hilbert polynomial of $C$. Set $H := \operatorname{Hilb}^{p(s)}(\mathbb{P}^{n+c})$. We will denote by $W(\underline{b}; \underline{a}) = W(b_1, \dots, b_t; a_0, a_1, \dots, a_{t+c-2}) \subset H$ the locus of good determinantal schemes $C \subset \mathbb{P}^{n+c}$ of codimension $c \geq 2$ defined by a homogeneous matrix $A = (f_{ij})$, where $i = 1, \dots t$, $j = 0, \dots, t+c-2$ and $f_{ij} \in k[x_0, \dots, x_{n+c}]$ is a homogeneous polynomial of degree $a_j - b_i$. We will bound $\dim W(\underline{b}; \underline{a})$ in terms of $b_1, \dots, b_t; a_0, a_1, \dots, a_{t+c-2}$ provided $(c-2)a_0 < \sum_{j=1}^{t+c-2} a_j - \sum_{i=1}^{t} b_i$.

To this end we consider the scheme $\mathbb{V} = \operatorname{Hom}_{\mathcal{O}_{\mathbb{P}^{n+c}}}(\mathcal{F}, \mathcal{G})$. It is an affine scheme of dimension $\sum_{i,j} h^0(\mathcal{O}_{\mathbb{P}^{n+c}}(a_i - b_j)) = \sum_{a_i \geq b_j} \binom{a_i - b_j + n + c}{n+c}$ whose rational points are the morphisms from $\mathcal{F}$ to $\mathcal{G}$. Let $\mathbb{Y}$ be the irreducible subscheme of

$\mathbb{V}$ whose rational points are the morphisms $\varphi_\lambda : \mathcal{F} \longrightarrow \mathcal{G}$ such that their associated homogeneous matrix $A_\lambda$ defines a good determinantal subscheme $C_\lambda \subset \mathbb{P}^{n+c}$. $\mathbb{Y} \subset \mathbb{V}$ is a non-empty open subscheme (indeed, $\varphi$ belongs to $\mathbb{Y}$).

Let $\mathbb{Y}^0 \subset \mathbb{Y}$ be the subscheme of $\mathbb{Y}$ whose rational points are the morphisms $\varphi_\lambda \in \mathbb{Y}$ such that $\operatorname{coker}(\varphi_\lambda) := \mathcal{B}_\lambda$ is a rank $c-1$ reflexive sheaf on $\mathbb{P}^{n+c}$ and let $\mathbb{Y}^0_S \subset \mathbb{Y}^0$ be the subscheme whose rational points are the morphisms $\varphi_\lambda \in \mathbb{Y}$ such that $\operatorname{coker}(\varphi_\lambda) := \mathcal{B}_\lambda$ is a rank $c-1$, stable, reflexive sheaf on $\mathbb{P}^{n+c}$.

LEMMA 10.1. (i) *$\mathbb{Y}^0 \subset \mathbb{Y}$ is a non-empty irreducible open subscheme.*

(ii) *If* $(c-2)a_0 < \sum_{j=1}^{t+c-2} a_j - \sum_{i=1}^{t} b_i$, *then $\mathbb{Y}^0_S \subset \mathbb{Y}$ is a non-empty irreducible open subscheme.*

PROOF. Since being stable and being reflexive are both open conditions, it is enough to prove that $\mathbb{Y}^0$ is non-empty and $\mathbb{Y}^0_S \subset \mathbb{Y}$ is non-empty provided $(c-2)a_0 < \sum_{j=1}^{t+c-2} a_j - \sum_{i=1}^{t} b_i$ . The result follows after a straightforward computation and we will first consider the case $c = 3$.

Codimension 3 case: We consider the exact sequence

$$0 \longrightarrow \mathcal{F} := \bigoplus_{i=1}^{t} \mathcal{O}_{\mathbb{P}^{n+3}}(b_i) \xrightarrow{A_\lambda} \mathcal{G} := \bigoplus_{j=0}^{t+1} \mathcal{O}_{\mathbb{P}^{n+3}}(a_j) \longrightarrow \mathcal{B}_\lambda \longrightarrow 0 \tag{10.1}$$

where $b_1 \geq ... \geq b_t$, $a_0 \geq a_1 \geq ... \geq a_{t+1}$ and

$$A_\lambda := \begin{pmatrix} x_0^{a_0-b_1} & x_1^{a_0-b_2} & x_2^{a_0-b_3} & 0 & 0 & \cdots & \cdots \\ 0 & x_0^{a_1-b_2} & x_1^{a_1-b_3} & x_2^{a_1-b_4} & 0 & 0 & \cdots \\ 0 & 0 & x_0^{a_2-b_3} & x_1^{a_2-b_4} & x_2^{a_2-b_5} & 0 & \cdots \\ \cdots & \cdots & \cdots & \cdots & \cdots & \cdots & \cdots \end{pmatrix}$$

Since $\operatorname{pd}(\mathcal{B}_\lambda) = 1$ and $\operatorname{codim}(S(\mathcal{B}_\lambda), \mathbb{P}^{n+3}) = 3$, where

$$S(\mathcal{B}_\lambda) = supp(\mathcal{E}xt^1(\mathcal{B}_\lambda, \mathcal{O}_{\mathbb{P}^{n+3}})),$$

we deduce that $\mathcal{B}_\lambda$ is a rank 2 reflexive sheaf on $\mathbb{P}^{n+3}$ ([**60**], Proposition 1.2). Moreover, $\mathcal{B}_\lambda$ is stable if and only if $H^0(\mathcal{B}_\lambda)_{norm} = 0$. But

$$c_1(\mathcal{B}_\lambda) = -\sum_{i=1}^{t} b_i + \sum_{j=0}^{t+1} a_j.$$

Therefore, $H^0(\mathcal{B}_\lambda)_{norm} = 0$ if and only if $H^0(\mathcal{B}_\lambda)(p) = 0$ for all

$$p \leq -\mu(\mathcal{B}_\lambda) = -c_1(\mathcal{B}_\lambda)/rk(\mathcal{B}_\lambda) = \left(\sum_{i=1}^{t} b_i - \sum_{j=0}^{t+1} a_j\right)/2$$

if and only if $a_0 < \sum_{i=1}^{t} b_i - \sum_{j=1}^{t+1} a_j$, where the last equivalence follows from the exact sequence (10.1).

Codimension $c \geq 4$ case: We consider the exact sequence

$$0 \longrightarrow \mathcal{F} := \bigoplus_{i=1}^{t} \mathcal{O}_{\mathbb{P}^{n+c}}(b_i) \xrightarrow{A_\lambda} \mathcal{G} := \bigoplus_{j=0}^{t+c-2} \mathcal{O}_{\mathbb{P}^{n+c}}(a_j) \longrightarrow \mathcal{B}_\lambda \longrightarrow 0 \tag{10.2}$$

where $b_1 \geq ... \geq b_t$, $a_0 \geq a_1 \geq ... \geq a_{t+c-2}$ and

$$A_\lambda := \begin{pmatrix} x_0^{a_0-b_1} & x_1^{a_0-b_2} & \cdots & x_{c-1}^{a_0-b_c} & 0 & 0 & \cdots & \cdots \\ 0 & x_0^{a_1-b_2} & x_1^{a_1-b_3} & \cdots & x_{c-1}^{a_1-b_{c+1}} & 0 & 0 & \cdots \\ 0 & 0 & x_0^{a_2-b_3} & x_1^{a_2-b_4} & \cdots & x_{c-1}^{a_2-b_{c+2}} & 0 & \cdots \\ \cdots & \cdots & \cdots & \cdots & \cdots & \cdots & \cdots & \cdots \end{pmatrix}$$

Again using [**60**] Proposition 1.2, together with the facts $\mathrm{pd}(\mathcal{B}_\lambda) = 1$ and

$$\mathrm{codim}(S(\mathcal{B}_\lambda), \mathbb{P}^{n+c}) \geq 3$$

we deduce that $\mathcal{B}_\lambda$ is a rank $c-1$ reflexive sheaf on $\mathbb{P}^{n+c}$. Let us check that $\mathcal{B}_\lambda$ is stable. Using the exact sequence (10.2) and the corresponding exact sequence involving alternating and symmetric powers,

$$0 \longrightarrow S^q\mathcal{F} \longrightarrow S^{q-1}\mathcal{F} \otimes \mathcal{G} \longrightarrow ... \longrightarrow \mathcal{F} \otimes \wedge^{q-1}\mathcal{G} \longrightarrow \wedge^q\mathcal{G} \longrightarrow \wedge^q\mathcal{B}_\lambda \longrightarrow 0,$$

we obtain $H^0(\wedge^l\mathcal{B}_\lambda)(m) = 0$ for all

$$m \leq -l\mu(\mathcal{B}_\lambda) = -\mu(\wedge^l\mathcal{B}_\lambda) = l(\sum_{i=1}^{t} b_i - \sum_{j=0}^{t+c-2} a_j)/(c-1).$$

Therefore, $H^0(\wedge^l\mathcal{B}_\lambda)_{norm} = 0$ for all $l$, $1 \leq l \leq c-2$; and $\mathcal{B}_\lambda$ is stable (Hoppe's criterion [**34**]), which proves Lemma 10.1. $\square$

The Eagon-Northcott complex of the universal morphism

$$\Psi : pr_2^*\mathcal{F} \longrightarrow pr_2^*\mathcal{G}$$

on $\mathbb{Y}^0 \times \mathbb{P}^{n+c}$ (where $pr_2 : \mathbb{Y}^0 \times \mathbb{P}^{n+c} \longrightarrow \mathbb{P}^{n+c}$ is the natural projection) induces a morphism

$$f : \mathbb{Y}^0 \longrightarrow W(\underline{b}; \underline{a})$$

which is defined by $f(\varphi_\lambda) := C_\lambda$ on closed points.

The affine group scheme $G := Aut(\mathcal{F}) \times Aut(\mathcal{G})$ operates on $\mathbb{Y}^0$:

$$\sigma : G \times \mathbb{Y}^0 \longrightarrow \mathbb{Y}^0; \quad ((\alpha, \beta), \varphi_\lambda) \mapsto \beta\varphi_\lambda\alpha^{-1}.$$

The action $\sigma$ is compatible with the morphism $f$, i.e. if two closed points $\varphi_{\lambda_1}$ and $\varphi_{\lambda_2}$ of $\mathbb{Y}^0$ belong to the same orbit of $\sigma$ then $f(\varphi_{\lambda_1}) = f(\varphi_{\lambda_2})$. Thus, at least set-theoretically $f : \mathbb{Y}^0 \longrightarrow W(\underline{b}, \underline{a})$ induces a surjective map from the orbit set $\mathbb{Y}^0//G$ to $W(\underline{b}, \underline{a})$. (It will be a bijection if $G$ acts transitively on the fibers of $f$.) Moreover, since the map from $\mathbb{Y}^0$ to the closure $\overline{W(\underline{b}, \underline{a})}$ in $\mathrm{Hilb}^{p(s)}(\mathbb{P}^{n+c})$ is dominant, we get that $W(\underline{b}, \underline{a})$ is irreducible, but not necessarily an irreducible component of $\mathrm{Hilb}^{p(s)}(\mathbb{P}^{n+c})$ (see Example 10.5 (4)).

Therefore, we have (small letters denote dimension):

$$\dim W(\underline{b}, \underline{a}) \leq hom_{\mathcal{O}_{\mathbb{P}^{n+c}}}(\mathcal{F}, \mathcal{G}) - aut(\mathcal{G}) - aut(\mathcal{F}) + \dim(G_\lambda) \tag{10.3}$$

where

$$\begin{aligned} G_\lambda &= \{(\delta, \tau) \in Aut(\mathcal{F}) \times Aut(\mathcal{G}) \mid \sigma((\delta, \tau), \varphi_\lambda) = \varphi_\lambda\} \\ &= \{(\delta, \tau) \in Aut(\mathcal{F}) \times Aut(\mathcal{G}) \mid \tau\varphi_\lambda\delta^{-1} = \varphi_\lambda\} \end{aligned}$$

is the isotropy group of a general closed point $\varphi_\lambda \in \mathbb{Y}^0$.

In next propositions we let $\binom{n+a}{n} = 0$ for $a < 0$, as usual.

PROPOSITION 10.2. (i) *For all* $\varphi_\lambda \in \mathbb{Y}^0$, *we have*

$$\dim(G_\lambda) = aut(\mathcal{B}_\lambda) + hom_{\mathcal{O}_{\mathbb{P}^{n+c}}}(\mathcal{G},\mathcal{F}) = aut(\mathcal{B}_\lambda) + \sum_{j,i}\binom{b_j - a_i + n + c}{n+c}.$$

(ii) *If* $(c-2)a_0 < \sum_{j=1}^{t+c-2} a_j - \sum_{i=1}^{t} b_i$, *then for all* $\varphi_\lambda \in \mathbb{Y}^0_S$, *we have*

$$\dim(G_\lambda) = 1 + hom_{\mathcal{O}_{\mathbb{P}^{n+c}}}(\mathcal{G},\mathcal{F}) = 1 + \sum_{j,i}\binom{b_j - a_i + n + c}{n+c}.$$

PROOF. (i) Since any automorphism $\beta \in Aut(\mathcal{B}_\lambda)$ lifts and determines automorphisms $\beta_{\mathcal{G}} \in Aut(\mathcal{G})$ and $\beta_{\mathcal{F}} \in Aut(\mathcal{F})$ such that the following diagram commutes:

$$\begin{array}{ccccccccc}
0 & \longrightarrow & \mathcal{F} & \xrightarrow{\varphi_\lambda} & \mathcal{G} & \longrightarrow & \mathcal{B}_\lambda & \longrightarrow & 0 \\
 & & \beta_{\mathcal{F}} \downarrow & & \downarrow \beta_{\mathcal{G}} & & \downarrow \beta & & \\
0 & \longrightarrow & \mathcal{F} & \xrightarrow{\varphi_\lambda} & \mathcal{G} & \longrightarrow & \mathcal{B}_\lambda & \longrightarrow & 0,
\end{array}$$

we see that $G_\lambda$ contains a non-empty open subset of $Aut(\mathcal{B}_\lambda) \times Hom_{\mathcal{O}_{\mathbb{P}^{n+c}}}(\mathcal{G},\mathcal{F})$. Indeed, for a general $\beta \in Aut(\mathcal{B}_\lambda)$ and a general $\rho \in Hom_{\mathcal{O}_{\mathbb{P}^{n+c}}}(\mathcal{G},\mathcal{F})$, the pair $(\delta,\tau) = (\beta_{\mathcal{F}} + \rho\varphi_\lambda, \beta_{\mathcal{G}} + \varphi_\lambda\rho)$ belongs to the isotropy group of $\varphi$ because

$$\begin{array}{rcl}
(\beta_{\mathcal{G}} + \varphi_\lambda\rho)\varphi_\lambda(\beta_{\mathcal{F}} + \rho\varphi_\lambda)^{-1} & = & (\beta_{\mathcal{G}}\varphi_\lambda + \varphi_\lambda\rho\varphi_\lambda)(\beta_{\mathcal{F}} + \rho\varphi_\lambda)^{-1} \\
 & = & (\varphi_\lambda\beta_{\mathcal{F}} + \varphi_\lambda\rho\varphi_\lambda)(\beta_{\mathcal{F}} + \rho\varphi_\lambda)^{-1} \\
 & = & \varphi_\lambda(\beta_{\mathcal{F}} + \rho\varphi_\lambda)(\beta_{\mathcal{F}} + \rho\varphi_\lambda)^{-1} \\
 & = & \varphi_\lambda.
\end{array}$$

Conversely, we claim that if $(\delta,\tau) \in G_\lambda$ then there exists $\beta \in Aut(\mathcal{B}_\lambda)$ and $\rho \in Hom_{\mathcal{O}_{\mathbb{P}^{n+c}}}(\mathcal{G},\mathcal{F})$ such that $(\delta,\tau) = (\beta_{\mathcal{F}} + \rho\varphi_\lambda, \beta_{\mathcal{G}} + \varphi_\lambda\rho)$. To this end, we consider the commutative diagram:

$$\begin{array}{ccccccccc}
0 & \longrightarrow & \mathcal{F} & \xrightarrow{\varphi_\lambda} & \mathcal{G} & \longrightarrow & \mathcal{B}_\lambda & \longrightarrow & 0 \\
 & & \delta \downarrow & & \downarrow \tau & & \downarrow \beta & & \\
0 & \longrightarrow & \mathcal{F} & \xrightarrow{\varphi_\lambda} & \mathcal{G} & \longrightarrow & \mathcal{B}_\lambda & \longrightarrow & 0
\end{array}$$

where we denote by $\beta \in Aut(\mathcal{B}_\lambda)$ the induced automorphism. Then $Im(\tau - \beta_{\mathcal{G}}) \subset Im(\varphi_\lambda)$ and thus there exists $\rho \in Hom_{\mathcal{O}_{\mathbb{P}^{n+c}}}(\mathcal{G},\mathcal{F})$ such that $\tau = \beta_{\mathcal{G}} + \varphi_\lambda\rho$. It follows that $\varphi_\lambda\delta = \tau\varphi_\lambda = (\beta_{\mathcal{G}} + \varphi_\lambda\rho)\varphi_\lambda = \varphi_\lambda(\beta_{\mathcal{F}} + \rho\varphi_\lambda)$, i.e., $\delta = \beta_{\mathcal{F}} + \rho\varphi_\lambda$. Hence, we have

$$\begin{array}{rcl}
\dim(G_\lambda) & = & aut(\mathcal{B}_\lambda) + hom_{\mathcal{O}_{\mathbb{P}^{n+c}}}(\mathcal{G},\mathcal{F}) \\
 & = & aut(\mathcal{B}_\lambda) + \sum_{i,j} h^0(\mathcal{O}_{\mathbb{P}^{n+c}}(b_j - a_i)) \\
 & = & aut(\mathcal{B}_\lambda) + \sum_{j,i}\binom{b_j - a_i + n + c}{n+c}.
\end{array}$$

(ii) If $(c-2)a_0 < \sum_{j=1}^{t+c-2} a_j - \sum_{i=1}^{t} b_i$, then for all $\varphi_\lambda \in \mathbb{Y}^0_S$, $\mathcal{B}_\lambda$ is stable. Hence, $\mathcal{B}_\lambda$ is simple, i.e., $Aut(\mathcal{B}_\lambda) = k$ and the result follows from (i). □

Finally from Proposition 10.2 and inequality (10.3) we easily deduce:

PROPOSITION 10.3. (i) *We have*

$$\dim W(\underline{b},\underline{a}) \leq \sum_{i,j}\binom{a_i - b_j + n + c}{n+c} - \sum_{i,j}\binom{a_i - a_j + n + c}{n+c} -$$

$$\sum_{i,j}\binom{b_i - b_j + n + c}{n+c} + \sum_{j,i}\binom{b_j - a_i + n + c}{n+c} + aut(\mathcal{B}_\lambda).$$

(ii) *If* $(c-2)a_0 < \sum_{j=1}^{t+c-2} a_j - \sum_{i=1}^{t} b_i$, *then*

$$\dim W(\underline{b},\underline{a}) \leq \sum_{i,j} \binom{a_i - b_j + n + c}{n+c} - \sum_{i,j} \binom{a_i - a_j + n + c}{n+c} -$$

$$\sum_{i,j} \binom{b_i - b_j + n + c}{n+c} + \sum_{j,i} \binom{b_j - a_i + n + c}{n+c} + 1.$$

We would like to know how far the bounds given in Proposition 10.3 are from being sharp, and to find $aut(\mathcal{B}_\lambda)$ in terms of $a_j$ and $b_i$ only. First of all, we will give examples where the hypothesis $(c-2)a_0 < \sum_{j=1}^{t+c-2} a_j - \sum_{i=1}^{t} b_i$ is satisfied.

REMARK 10.4. (1) Let $A = (f_{ij})_{i=1,\dots t}^{j=0,\dots,t+c-2}$ be a homogeneous matrix with linear entries. The inequality $(c-2)a_0 < \sum_{j=1}^{t+c-2} a_j - \sum_{i=1}^{t} b_i$ is clearly true.

(2) More generally, let $A = (f_{ij})_{i=1,\dots t}^{j=0,\dots,t+c-2}$ be a homogeneous matrix all whose entries are homogeneous polynomials of fixed degree d. Again, the inequality $(c-2)a_0 < \sum_{j=1}^{t+c-2} a_j - \sum_{i=1}^{t} b_i$ is verified.

(3) We consider a homogeneous $t \times (t+2)$ matrix $A = [L, M]$ where $M$ is a column of entries of degree $m$ and $L$ is a $t \times (t+1)$ matrix with linear entries (see Example 7.5). If $m < t+1$, then the condition $(c-2)a_0 < \sum_{j=1}^{t+c-2} a_j - \sum_{i=1}^{t} b_i$ is satisfied.

We give some examples for which the bound of Proposition 10.3 (ii) is sharp.

EXAMPLE 10.5. (1) According to [**22**], Theorem 2, in the codimension 2 case the bound is sharp.

(2) In the codimension 3 case, it follows from Proposition 10.7 and Corollary 10.15 that the inequality given in Proposition 10.3 (ii) is an equality provided a certain local complete intersection property holds.

(3) Let $C \subset \mathbb{P}^{n+4}$ be a good determinantal subscheme of codimension 4 defined by a $t \times (t+3)$ matrix with linear entries. Using the Macaulay program we have checked that for $(t,n) \in \{(2,3),(3,3)(2,4)\}$ the bound given in Proposition 10.3 (ii) is sharp.

(4) Let $C \subset \mathbb{P}^4$ be a good determinantal curve of degree 10 and arithmetic genus 6 defined by the maximal minors of a $3 \times 5$ matrix with linear entries. The closure of $W(\underline{b},\underline{a})$ inside $\mathrm{Hilb}^{10s-5}(\mathbb{P}^4)$ is not an irreducible component. Indeed, by Proposition 10.3 (ii), $\dim W(\underline{b},\underline{a}) \leq 42$ while the open set $H_{10,6,4}$ of $\mathrm{Hilb}^{10s-5}(\mathbb{P}^4)$ of smooth irreducible degree 10 and genus 6 curves in $\mathbb{P}^4$ is irreducible of dimension 45 (cf. [**20**], Theorem 8).

We are led to pose the following questions. When is the closure of $W(\underline{b},\underline{a})$ an irreducible component of $\mathrm{Hilb}^{p(s)}(\mathbb{P}^{n+c})$? Moreover, under which extra assumptions are the bounds given in Proposition 10.3 sharp?

For the remaining part of this chapter we concentrate on proving the unobstructedness, and finding the dimension of $\mathrm{Hilb}^p(\mathbb{P}^{n+c})$ and hence of $W(\underline{b};\underline{a})$, at a good determinantal scheme $C$ in $\mathbb{P}^{n+c}$ of codimension $c \geq 2$. By successively deleting columns (cf. first part of the proof of Theorem 3.6 or [**45**]), one may always suppose there is a "flag" of ACM subschemes

$$(\mathbf{X}.) : C = X_1 \subset X_2 \subset \dots \subset X_s \subset \mathbb{P}^{n+c} \tag{10.4}$$

where each $X_{i-1} \subset X_i$ (with ideal sheaf $\mathcal{I}_{X_{i-1}|X_i} = \mathcal{I}_i$) is of codimension 1, $X_s \subset \mathbb{P}^{n+c}$ is of codimension 2, and where there exist numbers $f_i$ and $\mathcal{O}_{X_i}$-modules $\mathcal{M}_i$ fitting into exact sequences

$$(10.5) \qquad 0 \to \mathcal{O}_{X_i}(-f_i) \to \mathcal{M}_i \to \mathcal{M}_{i-1} \to 0 \text{ for } 2 \le i \le s,$$

(which we obtain by twisting one of the exact sequences of Step II in Theorem 3.6) such that $\mathcal{I}_i(f_i)$ is the $\mathcal{O}_{X_i}$-dual of $\mathcal{M}_i$ for $2 \le i \le s$ and $\mathcal{M}_s = \omega_{X_s}$ is the canonical sheaf on $X_s$. Of course for $2 \le i \le s$ we let $I_i = \Gamma_*(X_i, \mathcal{I}_i)$ be the homogeneous ideal in $B_i = \Gamma_*(X_i, \mathcal{O}_{X_i}), M_i = \Gamma_*(X_i, \mathcal{M}_i)$ and $A = B_1 = B_2/I_2$. Since the projective dimension of $M_s$ is 2, the exact sequence above and the sequence

$$0 \to I_{X_i} \to I_{X_{i-1}} \to I_i \to 0$$

imply that the projective dimension of $M_{s-i}$ and of $I_{s-i}$ both equal $2+i$ (for $0 \le i \le s-2$) as $R$-modules.

REMARK 10.6. Let $D(q_1, \dots, q_s)$ be the Hilbert flag scheme parameterizing $s$-tuples $(\mathbf{X}.) = (X_1 \subset X_2 \subset \cdots \subset X_s \subset \mathbb{P}^{n+c})$ of equidimensional, locally Cohen-Macaulay closed subschemes in $\mathbb{P}^{n+c}$, where $q_i$ is the Hilbert polynomial of $X_i$. Note that, given $(\mathbf{X}.) \in D(q_1, \dots, q_s)$ and the short exact sequence of the $\mathcal{M}_i$'s satisfying the set-up of (10.4) and (10.5) above, then one may get back a matrix whose maximal minors define the good determinantal ACM subscheme $C$, by using (3.1). Moreover, supposing (10.12), we will later see that given any flat deformation $(\mathbf{X}._T)$ of $(\mathbf{X}.)$ (cf. the proof of Proposition 10.12 and let $C_T = X_{1,T}$), then the corresponding short exact sequence of the $\Gamma_*(\mathcal{M}_{i,T})$'s and the diagram (3.1) show the existence of a matrix over $T$ (reducing to the matrix which defines $C$) whose associated Eagon-Northcott complex defines a minimal resolution of $\mathcal{O}_{C_T}$, because a complex which lifts a minimal resolution is again a resolution. This shows that the closure of $W(\underline{b}; \underline{a})$ in $\mathrm{Hilb}^p(\mathbb{P}^{n+c})$ coincides with the closure of the image $pr_1(W(\mathbf{X}.))$ (where $pr_1 : D(q_1, \dots, q_s) \to \mathrm{Hilb}^p(\mathbb{P}^{n+c})$ is the first projection) of the irreducible component $W(\mathbf{X}.) \subseteq D(q_1, \dots, q_s)$ to which $(\mathbf{X}.)$ belongs. Note that $W(\mathbf{X}.)$ is unique by the unobstructedness of Proposition 10.12. We return to a more detailed discussion of these facts in Remark 10.14.

Now if all $X_{i-1} \subset X_i$ are Cartier divisors (an assumption which we later will weaken), the set-up of (10.4) fits nicely with the theory developed in Chapter 9. Indeed, by Remark 10.6, if $(\mathbf{X}.) = (C \subset S \subset \mathbb{P}^{n+3})$, i.e. $s = 2$ and $c = 3$ and $S$ satisfies $G_1$ (to fulfill (10.12) as required by Remark 10.6), then the closure of $W(\underline{b}; \underline{a})$ is precisely what we called $W(C \subset S)$ in Proposition 9.14. Under the assumption of Theorem 9.4, $C$ was unobstructed and $W(C \subset S)$ was an irreducible component. We can therefore use these results to prove that $C$ is unobstructed, and to decide when $W(\underline{b}; \underline{a})$ is an irreducible component and to compute $\dim W(\underline{b}; \underline{a})$. We remark, however, that the unobstructedness of most generic determinantal schemes of any codimension follows from works of Svanes ([**11**],Theorem 15.10). Indeed, in this generic case, $H^2(R, A, A) \cong \mathrm{Ext}^2_A(\Omega_A, A) = 0$ where $\Omega_A$ is the module of differentials. If $c = 3$ and $n \le 2$, however, then $H^2(R, A, A)$ is in general far from being vanishing (cf. Example 7.5), and in any case it is interesting to find $\dim W(\underline{b}; \underline{a})$ and to prove unobstructedness more generally.

When we consider good determinantal schemes, we can without loss of generality suppose that each homogeneous entry of the matrix of degree less than 1 is zero and that no column or row consists entirely of zeros.

PROPOSITION 10.7. *Let $C \subset S$ be good determinantal schemes in $\mathbb{P}^{n+3}$ where $C$ is given by the maximal minors of a $t \times (t+2)$ matrix whose $ij$-th entry is homogeneous of degree $a_j - b_i$, $0 \le j \le t+1$, $1 \le i \le t$, and $S$ is correspondingly given by deleting some column. Let*

$$l = \sum_{j=0}^{t+1} a_j - \sum_{i=1}^{t} b_i, \quad a_r = \max\{a_j\}$$

*and suppose $C \subset S$ is Cartier, $S$ satisfies $G_1$ and $n = \dim C \ge 1$. Then*

(i) $\dim W(\underline{b}; \underline{a}) = \lambda + \kappa$ *where*

$$\begin{aligned} \lambda &= \sum_{i,j} \binom{a_j - b_i + n + 3}{n+3} + \sum_{i,j} \binom{b_i - a_j + n + 3}{n+3} - \sum_{i,j} \binom{b_i - b_j + n + 3}{n+3} \\ &\quad - \sum_{i,j} \binom{a_j - a_i + n + 3}{n+3} + 1 \\ \kappa &= \binom{-l + 2a_r + n + 3}{n+3}. \end{aligned}$$

(ii) *Suppose that either of the following conditions holds:*
   a. $\dim C > 1$, *or*
   b. $\dim C = 1$ *and* $H^1(\mathcal{I}_S/\mathcal{I}_S^2(l - 2a_k)) = 0$, *where $S$ is obtained deleting the $(k+1)$-st column (say).*

*Then $C$ is unobstructed, and*

$$\dim_C \mathrm{Hilb}^p(\mathbb{P}^{n+3}) = \dim W(\underline{b}; \underline{a}) = \lambda + \kappa.$$

REMARK 10.8. It follows from Proposition 10.7 that, in the codimension 3 case, the bounds given in Proposition 10.3 (with $\kappa$ instead of $aut(\mathcal{B}_\lambda) - 1$) are sharp. So, in the codimension 3 case, we have a dimension formula of the strata of $\mathrm{Hilb}^p(\mathbb{P}^{n+3})$ which corresponds to good determinantal schemes, and we are led to ask whether for higher codimension the bounds given in Proposition 10.3 are also sharp.

PROOF. Since the proof of (i) requires arguments from the proof of the unobstructedness of $C$ (we need $\alpha_L = 0$), we prove the unobstructedness first.

Note that $\omega_S \cong \mathcal{O}_S(C)(-f)$ for some $f$ (which we later will calculate and we get $f = n + 4 + 2a_0 - l$) as in the set-up of (10.4) and (10.5). Hence we have

$$\begin{aligned} \mathcal{N}_S \otimes \mathcal{I}_{C|S} &\cong \mathcal{E}xt^1(\mathcal{I}_S, \mathcal{I}_S) \otimes \mathcal{O}_S(-C) \\ &\cong \mathcal{E}xt^1(\mathcal{I}_S, \mathcal{O}_{\mathbb{P}}) \otimes \mathcal{I}_S \otimes \mathcal{O}_S(-C) \\ &\cong \mathcal{I}_S \otimes \mathcal{O}_S(n + 4 - f) \end{aligned}$$

because

$$\mathcal{E}xt^1(\mathcal{I}_S, \mathcal{O}_{\mathbb{P}}(-n-4)) \cong \mathcal{E}xt^2(\mathcal{O}_S, \mathcal{O}_{\mathbb{P}}(-n-4)) \cong \omega_S \cong \mathcal{O}_S(C)(-f),$$

i.e. the group $H^1(\mathcal{N}_S \otimes \mathcal{I}_{C|S})$ of Theorem 9.4 and $H^1(\mathcal{I}_S/\mathcal{I}_S^2(n+4-f))$ above coincide. We will use Theorem 9.4 to prove that $C$ is unobstructed provided $H^1(\mathcal{N}_S \otimes \mathcal{I}_{C|S}) = 0$ and that this $H^1$- group always vanishes if $\dim C > 1$.

To prove it, we first claim that the relative Picard scheme $Pic$ is formally smooth at $(\mathcal{O}_S(C), S)$. Indeed if $\dim S > 2$ we get $H^2(\mathcal{O}_S) = 0$ because $S$ is an ACM scheme, and we conclude by Proposition 9.2. In the case $\dim S = 2$ we have $H^1(\mathcal{O}_S) = 0$ by the ACM property of $S$. It follows that any deformation of $\mathcal{O}_S(C)$ to a given deformation of $S$ is unique up to isomorphism [**27**]. To prove that $Pic$ is formally smooth at $(\mathcal{O}_S(C), S)$ it suffices to take an arbitrary (flat) deformation $(L_T, S_T)$ of $(\mathcal{O}_S(C), S)$ to a local artinian $k$-algebra $T$ with residue field $k$ and to deform it further to any local artinian $T'$ having $T$ as a quotient. Now since $S$ is of codimension 2 in $\mathbb{P}^{n+3}$, we know that $S$ is unobstructed by [**22**] or Proposition 6.17, i.e. a deformation $S_{T'}$ of $S_T$ exists. Moreover since $\mathcal{O}_S(C)(-f)$ is the canonical sheaf $\mathcal{E}xt^2(\mathcal{O}_S, \mathcal{O}_{\mathbb{P}}(-n-4))$ and $\mathcal{E}xt^i(\mathcal{O}_S, \mathcal{O}_{\mathbb{P}}(-n-4)) = 0$ for $i \neq 2$, one proves that $\mathcal{E}xt^2(\mathcal{O}_{S_T}, \mathcal{O}_{\mathbb{P}_T}(-n-4))$ is a flat deformation of $\mathcal{O}_S(C)(-f)$ (as in [**38**], first part of the proof of Proposition 2.4, where we used the vanishing of $\mathrm{Ext}^1$-groups to see that the corresponding Hom deforms flatly), or see [**18**], Proposition (A1). By uniqueness, this $\mathcal{E}xt^2$ group is isomorphic to $L_T(-f)$ which we similarly can deform to $\mathcal{E}xt^2(\mathcal{O}_{S_{T'}}, \mathcal{O}_{\mathbb{P}_{T'}}(-n-4))$ over $T'$, and the claim is proved.

Moreover if $\dim C > 1$, we claim that $H^1(\mathcal{N}_S \otimes \mathcal{I}_{C|S}) = 0$. Indeed since $S$ is generically a complete intersection [**45**] and an ACM subscheme of codimension 2 in $\mathbb{P}^{n+3}$, one knows that the projective dimension of $\mathcal{I}_S/\mathcal{I}_S^2 \leq 3$ by [**3**], Theorem 3.2 (cf. Remark 10.9). If we for instance split the projective resolution of $\mathcal{I}_S/\mathcal{I}_S^2$ into short exact sequences and take cohomology, we get

$$H^1(\mathcal{N}_S \otimes \mathcal{I}_{C|S}) \cong H^1(\mathcal{I}_S/\mathcal{I}_S^2(n+4-f)) = 0.$$

All assumptions of Theorem 9.4 are therefore satisfied. Indeed,

$$H^1(\mathcal{O}_S(C)) = H^1(\omega_S(f)) = 0$$

by duality, and since $H^1(\mathcal{N}_S \otimes \mathcal{I}_{C|S}) = 0$ (by assumption if $n = 1$), we find $C$ to be unobstructed by Theorem 9.4.

To prove the dimension formulas, we need a closer look at the proof of the first claim above in the case $\dim S = 2$. Letting $T$ be $k$ and $T'$ be dual numbers $k[\epsilon]$, we see that to every deformation $S_{k[\epsilon]}$ of $S$ to $k[\epsilon]$ there exists a deformation of $L = \mathcal{O}_S(C)$ to $S_{k[\epsilon]}$. By the description of the map $\alpha_L : H^0(\mathcal{N}_S) \to H^2(\mathcal{O}_S)$ appearing in the proof of Theorem 9.4, we get $\alpha_L = 0$. By Theorem 9.4 it follows that

(10.6)

$$\dim_C \mathrm{Hilb}^p(\mathbb{P}^{n+3}) = h^0(\mathcal{N}_S) - \dim(I_S)_{n+4-f} + \dim(I_S^2)_{n+4-f} + h^0(\omega_S(f)) - 1$$

in which we have used that the projective dimension of $I_S/I_S^2$ is at most 3 to see the first isomorphism in

$$I_S/I_S^2 \cong H^0_*(\mathcal{I}_S/\mathcal{I}_S^2) \cong H^0_*(\mathcal{N}_S \otimes \mathcal{I}_{C|S})(f-n-4).$$

By Remark 10.6, $\overline{W(\underline{b};\underline{a})} = \overline{pr_1(W(\mathbf{X}.))}$. Since the proof of Proposition 9.2 (iii) shows that the vanishing of $H^1(\mathcal{N}_S \otimes \mathcal{I}_{C|S})$ implies the smoothness of $pr_1 : D(p,q) \longrightarrow \mathrm{Hilb}^p(\mathbb{P}^{n+3})$ at $(\mathbf{X}.) = (C \subset S \subset \mathbb{P}^{n+3})$, we get

$$\dim_C \mathrm{Hilb}^p(\mathbb{P}^{n+3}) = \dim W(\underline{b};\underline{a}).$$

If the assumption $H^1(\mathcal{N}_S \otimes \mathcal{I}_{C|S}) = 0$ is not satisfied, then Proposition 9.14 shows that the right hand side of (10.6) is equal to $\dim W(C \subset S) = \dim pr_1(W(\mathbf{X}.))$, i.e. to $\dim W(\underline{b};\underline{a})$ because Pic is formally smooth at $(\mathcal{O}_S(C), S)$ by the first proven claim of this proof. Hence (i) and (ii) follow if we can show that the right hand side of (10.6) is $\lambda + \kappa$.

Without loss of generality we can suppose that $S$ is obtained by deleting the first column. Then $C$ and $S$ are the degenerecy loci of

$$\bigoplus_{i=1}^{t} R(b_i) \to \bigoplus_{j=0}^{t+1} R(a_j) \quad \text{and} \quad \bigoplus_{i=1}^{t} R(b_i) \to \bigoplus_{j=1}^{t+1} R(a_j)$$

respectively, leading to the minimal resolutions

$$0 \to F \to G \to I_S \to 0 \tag{10.7}$$

$$0 \to R \to G^* \to F^* \to K_B(n+4) \to 0. \tag{10.8}$$

Note that $K_B$ is the canonical module of $S$, and

$$F = \bigoplus_{i=1}^{t} R(b_i + a_0 - l), \qquad G = \bigoplus_{j=1}^{t+1} R(a_j + a_0 - l). \tag{10.9}$$

As in the proof of Theorem 3.6 there is an exact sequence (cf. (10.5))

$$0 \to R/I_S(-2a_0 + l) \to K_B(n+4) \to M_C(n+4) \to 0$$

and we get $\omega_S(n+4+2a_0-l) \cong \mathcal{O}_S(C)$, i.e. $f = n+4+2a_0-l$, using that the $G_1$ property implies the reflexivity of $\omega_S$. By using (10.7) and (10.8) we have

$$\dim(I_S)_{n+4-f} = \sum_{j=1}^{t+1} \binom{a_j - a_0 + n + 3}{n+3} - \sum_{i=1}^{t} \binom{b_i - a_0 + n + 3}{n+3}$$

$$h^0(\omega_S(f)) = \sum_{i=1}^{t} \binom{a_0 - b_i + n + 3}{n+3} - \sum_{j=1}^{t+1} \binom{a_0 - a_j + n + 3}{n+3} + \binom{-l + 2a_0 + n + 3}{n+3}$$

which one may combine with the dimension formula of $h^0(\mathcal{N}_S)$ given in [**22**];

$$h^0(\mathcal{N}_S) = \sum_{1 \le i,j} \binom{a_j - b_i + n + 3}{n+3} + \sum_{1 \le i,j} \binom{b_i - a_j + n + 3}{n+3}$$
$$- \sum_{i,j} \binom{b_i - b_j + n + 3}{n+3} - \sum_{1 \le i,j} \binom{a_i - a_j + n + 3}{n+3} + 1.$$

This leads exactly to

$$h^0(\mathcal{N}_S) - \dim(I_S)_{n+4-f} + h^0(\omega_S(f)) - 1 = \lambda + \binom{-l + 2a_0 + n + 3}{n+3}.$$

Finally to compute $\dim(I_S^2)_{n+4-f}$, we have by (10.7) and Remark 10.9 an exact sequence

$$0 \to \wedge^2(F) \to F \otimes G \to S^2(G) \to I_S^2 \to 0.$$

Since $n+4-f = l - 2a_0$, then

$$\bigoplus_{1 \le i,j} R(-l + a_i + b_j)_0 \to \bigoplus_{1 \le i \le j \le t+1} R(-l + a_i + a_j)_0 \to (I_S^2)_{n+4-f} \to 0 \tag{10.10}$$

is exact. By the Eagon-Northcott resolution of $I_C$, we see that the minimal generators of $I_C$ have degree $l - a_i - a_j$ for $i < j$, i.e.

$$R(-l + a_i + a_j)_0 = 0 \text{ for } 0 \le i < j \le t+1.$$

Now if $r$ is the smallest integer such that $a_r = \max\{a_i\}$, then (10.10) reduces to

$$R(-l + 2a_r)_0 \cong (I_S^2)_{n+4-f} \quad \text{provided } r \ge 1$$

because, for $k \ne r$, we have $a_r \ge a_k$ and

$$0 > -l + a_r + a_k \ge -l + 2a_k. \tag{10.11}$$

From (10.11) we also get $l > 2a_0$ if $r \geq 1$ and $l > 2a_k$ (with $k > 0$) if $r = 0$. Hence

$$\lambda + \binom{-l+2a_0+n+3}{n+3} + \dim(I_S^2)_{n+4-f} = \lambda + \binom{-l+2a_r+n+3}{n+3}$$

in any case, and the proof is complete. □

REMARK 10.9. With notations and assumptions as in Proposition 10.7 (i), we easily see from the proof that

$$\mathcal{O}_S(C) \cong \omega_S(f), \text{ and}$$
$$H^1(\mathcal{N}_S \otimes \mathcal{I}_{C|S}) \cong H^1(\mathcal{I}_S/\mathcal{I}_S^2(n+4-f))$$

where $f = n+4+2a_k - l$ provided $S$ is obtained by deleting the $(k+1)$-st column. The vanishing of $H^1(\mathcal{I}_S/\mathcal{I}_S^2(n+4-f))$ is therefore equivalent to the exactness of (7.2). Moreover invoking Proposition 9.14 (and Remark (10.6)), we see that

$$\dim_C \operatorname{Hilb}^p(\mathbb{P}^{n+3}) \leq \dim W(\underline{b};\underline{a}) + h^1(\mathcal{I}_S/\mathcal{I}_S^2(l-2a_k)).$$

So, it is interesting to compute $h^1(\mathcal{I}_S/\mathcal{I}_S^2(l-2a_k))$. Using results of [**3**] we can at least partially find the dimension. Indeed if

$$0 \to F \to G \to I_S \to 0$$

is a minimal resolution and $\mathbf{K}.$ is the Koszul complex of a set of minimal generators of $I_S$ (so $K_1 = G$) and $H_1$ its first homology, then the sequences

$$0 \to H_1 \to K_1 \otimes_R B \to I_S/I_S^2 \to 0, \text{ and}$$
$$0 \to \wedge^2(F) \to \wedge^2(G) \to F \to H_1 \to 0$$

are exact (cf. [**3**], Lemma 1.6). Note that there is another related exact sequence [**70**]

$$0 \to \wedge^2(F) \to F \otimes G \to S^2(G) \to I_S^2 \to 0$$

with which we can make equivalent computations.

As a special case, we reconsider Example 10.5 (4). Since we can continue (10.10) to the left by

$$0 \to (\wedge^2 F)_{n+4-f} \to \cdots$$

where $n+4-f = l - 2a_0$, we get

$$h^1(\mathcal{I}_S/\mathcal{I}_S^2(l-2a_k)) = h^2(\mathcal{I}_S^2(3)) \leq h^4(\wedge^2\tilde{F}(3)) = 3$$

in accordance with what we found in Example 10.5 (4).

EXAMPLE 10.10. Let $C$ be a good determinantal subscheme of $\mathbb{P}^{n+3}$ given by the maximal minors of a $t \times (t+2)$ matrix $[M, L]$ where $M$ is a column of entries of degree $m$ and $L$ is a matrix of linear entries whose minors define a subscheme $S$ (cf. Example 7.5). To apply Proposition 10.7 we suppose $C \subset S$ is a Cartier divisor, $S$ satisfies $G_1$, and $n \leq 2$. By Remark 10.9 we know the vanishing of $H^1(\mathcal{I}_S/\mathcal{I}_S^2(l-2a_0))$ is equivalent to the exactness of the complex (7.2). Since the calculations of Example 7.5 show that the middle term of (7.2) vanishes if $m \geq 3$, we get $H^1(\mathcal{I}_S/\mathcal{I}_S^2(l-2a_0)) = 0$. So if $\dim C > 1$ or if $\dim C = 1$ and $m \geq 3$, then $C$ is unobstructed and

$$\begin{aligned}\dim_C \operatorname{Hilb}^p(\mathbb{P}^{n+3}) \;=\; & \binom{m+n+3}{n+3} t + t\,(t+1)\,(n+2) - 1 \\ & -(t+1)\binom{m+n+2}{n+3} + \binom{m-t+n+2}{n+3} \text{ if } m > 1\end{aligned}$$

and

$$\dim_C \operatorname{Hilb}^p(\mathbb{P}^{n+3}) \quad = \quad (n+4)\, t\, (t+2) - 2t^2 - 4t - 3, \text{ if } m = 1.$$

COROLLARY 10.11. *Let $C \subset S$ be the good determinantal schemes in $\mathbb{P}^{n+3}$ of Proposition 10.7, where $S$ is obtained by deleting the first column, let $b_1 \geq \cdots \geq b_t$ and $a_0 \geq a_1 \geq \cdots \geq a_t \geq a_{t+1}$, and suppose $a_0 > a_1 + a_2 - b_t$ and $\dim C = 1$. Then $H^1(\mathcal{I}_S/\mathcal{I}_S^2(l - 2a_0)) = 0$, i.e. $C$ is unobstructed and $\dim_C \operatorname{Hilb}^p(\mathbb{P}^{n+3}) = \dim W(\underline{b}; \underline{a}) = \lambda + \kappa$.*

PROOF. By Remark 10.9, $H^1(\mathcal{I}_S/\mathcal{I}_S^2(l - 2a_0)) = 0$ provided the middle term of (7.2), namely $\bigoplus_{i=1}^t H^0(\mathcal{I}_{C|S}(l - a_0 - b_i))$, vanishes (cf. (10.9)). Using this, we get the corollary by making the minimal generators of $I_{C|S}$ explicit. Indeed since the maximal minors of the $t \times (t+2)$ matrix of Proposition 10.7, which are not minimal generators of $I_S$, have degree $l - a_j - a_k$ for $1 \leq j < k \leq t+1$, we get the result from Proposition 10.7 because the assumption $a_0 > a_1 + a_2 - b_t$ implies $l - a_j - a_k > l - a_0 - b_i$ for any $i = 1, ..., t$. □

In the following we will generalize the considerations of Proposition 10.7 to determinantal subschemes of higher dimension and codimension. Since it is for instance well-known that the singular locus of a generic determinantal codimension 2 subscheme sits in codimension 6 in $\mathbb{P}^{n+c}$, the Cartier assumption of Proposition 10.7 should be weakened if we satisfactorily want to treat higher dimensional determinantal schemes. For good determinantal schemes (Definition 2.1), however, the locus of non-vanishing submaximal minors intersect properly each component. Consulting [**45**] one may see there always is a flag as in (10.4) which furthermore satisfies:

(10.12)

For each $i > 1$, the closed embeddings $X_i \subset \mathbb{P}^{n+c}$ and $X_{i-1} \subset X_i$ are local complete intersections outside some set $Z_i$ of codimension 1 in $X_{i-1}$ ($\operatorname{depth}_{Z_i} \mathcal{O}_{X_{i-1}} \geq 1$).

This condition guarantees that the $\mathcal{O}_{X_i}$-modules $\mathcal{M}_i$ and $\mathcal{I}_i(f_i)$ are determined by their sections of $U_i = X_i - Z_i$, that they are reflexive and dual to each other. Indeed since we have $pd\, M_{s-i} = pd\, I_{s-i} = 2+i$, we get $\operatorname{depth}_{I_{Z_i}} M_i \geq 2$ and $\operatorname{depth}_{I_{Z_i}} I_i \geq 2$ [**68**], Lemma 3.5, and since $\operatorname{depth}_{I_{Z_i}} B_i \geq 2$ we easily deduce $\operatorname{depth}_{I_{Z_i}} M_i^* \geq 2$ and $\operatorname{depth}_{I_{Z_i}} I_i^* \geq 2$ for their duals as well. Hence we conclude by the invertibility of $\mathcal{M}_i$ and $\mathcal{I}_i$ restricted to $U_i$ and by

$$(10.13) \qquad \begin{array}{ll} M_i = \Gamma_*(X_i, \mathcal{M}_i) \cong \Gamma_*(U_i, \mathcal{M}_i) & \text{and} \quad \Gamma_*(X_i, \mathcal{M}_i^*) \cong \Gamma_*(U_i, \mathcal{M}_i^*) \\ I_i = \Gamma_*(X_i, \mathcal{I}_i) \cong \Gamma_*(U_i, \mathcal{I}_i) & \text{and} \quad \Gamma_*(X_i, \mathcal{I}_i^*) \cong \Gamma_*(U_i, \mathcal{I}_i^*). \end{array}$$

Note that (10.13) follows from the vanishing of the corresponding local cohomology sheaves (cf. [**29**]), from which we also see that (10.13) holds if we intersect the sets $X_i$ and $U_i$ by any open set $U$ of $X_i$. Moreover note that when we consider flat deformations of a determinantal flag (10.4), the projective dimension of $I_i$, hence the inequalities of the depth conditions of $I_i$ and $I_i^*$ above, remains unchanged, i.e. at least the lower line of (10.13) holds for the corresponding deformations, allowing us to prove the following technical result

PROPOSITION 10.12. *Let $C = X_1 \subset X_2 \subset \cdots \subset X_s \subset \mathbb{P}^{n+c}$ be a "flag" ($\mathbf{X}$.) of determinantal ACM subschemes as in (10.4) and (10.12), and let*

$$R \to B_s \to \cdots \to B_2 \to A$$

*be the corresponding "flag"* (**B**.) *of determinantal graded Cohen-Macaulay quotients. Then* (**B**.) *is unobstructed as a flag. Moreover if* $\dim C \geq 1$*, then* (**X**.) *is unobstructed as a flag, i.e. to any (flat) deformation*

$$(\mathbf{X}_{.,T}) : C_T = X_{1,T} \subset \cdots \subset X_{s,T} \subset \mathbb{P}_T^{n+c}$$

*of* (**X**.) *to an arbitrary local artinian* $k$*-algebra* $T$ *with residue field* $k$*, there exists a deformation of* $(\mathbf{X}_{.,T})$ *further to any local artinian* $T'$ *having* $T$ *as a quotient.*

PROOF. We only prove that (**X**.) is unobstructed because (**B**.) is seen to be unobstructed by almost the same proof. Now, since a flat deformation $(\mathbf{X}_{.,T})$ is given, the sheaf ideals $\mathcal{I}_{i,T}$ of $X_{i-1,T} \subset X_{i,T}$ are flat over $T$, and, as we remarked above, satisfy (10.13) (obviously modified by indexing by $T$). Moreover the canonical sheaf $\mathcal{E}xt^2(\mathcal{O}_{X_s,T}, \mathcal{O}_{P_T}(-n-4))$ is $T$-flat because the $\mathcal{E}xt^i$-groups vanish for $i \neq 2$ (cf. [**18**], Proposition (A1)). The same argument shows that the corresponding graded $\mathrm{Ext}^2$-group is $T$-flat because $\Gamma_*(\mathcal{O}_{X_s,T})$ is a flat deformation of $\Gamma_*(\mathcal{O}_{X_s})$ by the ACM-property of $X_s$, cf.[**22**]. Letting $M_{s,T}$ be this graded $\mathrm{Ext}^2$-group by definition, then $M_{s,T}$ is $T$-flat and it maps surjectively onto $M_s \cong \mathrm{Ext}^2(B_s, R(-n-4))$ via $\otimes_T k$. Hence there exists a $T$-flat $\mathcal{M}_{s,T}$ and a well defined map $B_{s,T}(-f_s) \to M_{s,T}$ whose corresponding section $\mathcal{O}_{X_s,T}(-f_s) \to \mathcal{M}_{s,T}$ commutes with $\mathcal{O}_{X_s}(-f_s) \to \mathcal{M}_s$ . Defining successively $M_{i,T}$ by the sequence

$$0 \to B_{i+1,T}(-f_{i+1}) \to M_{i+1,T} \to M_{i,T} \to 0, \tag{10.14}$$

it follows that $M_{i,T}$ are $T$-flat for any $i \geq 2$ and satisfy (10.13). Again, since $M_{i,T}$ maps surjectively onto $M_i$, the section $\mathcal{O}_{X_i}(-f_i) \to \mathcal{M}_i$ lifts to a section of $\mathcal{O}_{X_i,T}(-f_i) \to \mathcal{M}_{i,T}$.

After having made an appropriate modification of the choice of the liftings of these sections (described below) and hence of their cokernels

$$M_{i,T} = \mathrm{coker}(B_{i+1,T}(-f_{i+1}) \to M_{i+1,T}),$$

we claim that $\mathcal{I}_{i,T}(f_i)$ is the $\mathcal{O}_{X_i,T}$-dual of $\mathcal{M}_{i,T}$ for $i \geq 2$. Suppose the claim holds for subscripts greater than $i$. Since $\mathcal{M}_{i,T}^*$ has depth at least 2 (because $B_{i,T}$ has), i.e. (10.13) indexed by $T$ holds for $M_{i,T}^*$, we see that $\mathcal{M}_{i,T}^*(-f_i)$ is a $T$-flat sheaf of ideals. The two flat ideals $\mathcal{M}_{i,T}^*(-f_i)$ and $\mathcal{I}_{i,T}$ correspond to two deformations of $X_{i-1} \subset X_i$ to $X_{i,T}$, i.e "their difference" defines an element $\gamma$ of $H^0(\mathcal{N}_{X_{i-1}|X_i})$ (if we successively decompose $T \to k$ into small surjections $T_j \to T_{j-1}$ for which we can suppose ker $(T_j \to T_{j-1}) \cong k$). Now since $\mathcal{N}_{X_{i-1}|X_i} = \mathcal{H}om(\mathcal{I}_i, \mathcal{O}_{X_{i-1}})$ and $\mathcal{M}_i(f_i) = \mathcal{H}om(\mathcal{I}_i, \mathcal{O}_{X_i})$ there exists an exact sequence

$$0 \to \mathcal{O}_{X_i} \to \mathcal{M}_i(f_i) \to \mathcal{N}_{X_{i-1}|X_i} \to 0, \tag{10.15}$$

which in view of the vanishing of $H^1(\mathcal{O}_{X_i})$, defines an element of $H^0(\mathcal{M}_i(f_i))$ which maps onto $\gamma$ in $H^0(\mathcal{N}_{X_{i-1}|X_i})$. This element precisely modifies the choice of the lifted section of $\mathcal{O}_{X_{i,T}}(-f_i) \to \mathcal{M}_{i,T}$ so that the claim holds.

Now we can easily deform $(\mathbf{X}_{.,T})$ further to $T'$. Indeed $R_T \to B_{s,T}$ , and hence $\mathrm{Ext}^2(B_{s,T}, R_T(-n-4))^\sim$ deforms further to $T'$ because in codimension 2, we know that $R \to B_s$ is unobstructed by Proposition 6.17. Hence we get a deformation of $\mathcal{O}_{X_{s,T}}(-f_s) \to \mathcal{M}_{s,T} \cong \mathrm{Ext}^2(B_{s,T}, R_T(-n-4))^\sim$ to $T'$. If we successively define $M_{i,T'}$ for $i \geq 2$, by the sequence

$$0 \to B_{i+1,T'}(-f_{i+1}) \to M_{i+1,T'} \to M_{i,T'} \to 0,$$

it follows that $\mathcal{M}_{i,T'}$ are $T'$- flat and that $\Gamma_*(\mathcal{M}_{i,T'})$ is a flat deformation of (and hence maps surjectively onto) $\Gamma_*(\mathcal{M}_{i,T})$. We therefore get a section of $\mathcal{M}_{i,T'}(f_i)$ whose dual is a flat sheaf ideal defining $X_{i-1,T'}$ in $X_{i,T'}$. By the proven claim above, $X_{i-1,T'} \subset X_{i,T'}$ becomes $X_{i-1,T} \subset X_{i,T}$ when we tensorize (i.e. take the pullback) with $T$ over $T'$ and the proof is complete. $\square$

Now we are able to prove a main result of this chapter.

THEOREM 10.13. *Let $C \subset \mathbb{P}^{n+c}$ be a good determinantal scheme of dimension $n = \dim C \geq 1$ (resp. $\dim C = 0$). If $C = X_1 \subset X_2 \subset \cdots \subset X_s \subset \mathbb{P}^{n+c}$ is a flag as in (10.4) and (10.12) for which the natural maps*

$$\operatorname{Ext}^1_{\mathcal{O}_P}(\mathcal{I}_{X_i}, \mathcal{I}_i) \to \operatorname{Ext}^1_{\mathcal{O}_P}(\mathcal{I}_{X_i}, \mathcal{O}_{X_i}) \quad (\text{resp. } {}_0\operatorname{Ext}^1_R(I_{X_i}, I_i) \to {}_0\operatorname{Ext}^1_R(I_{X_i}, B_i))$$

*are injective for any $i = 2, 3, ..., s$, then $C$ is unobstructed (resp. $H$-unobstructed). Moreover*

$$\begin{aligned}\dim_C \operatorname{Hilb}^p(\mathbb{P}^{n+c}) &= h^0(X_s, \mathcal{N}_{X_s}) + \sum_{i=2}^{s} h^0(X_{i-1}, \mathcal{N}_{X_{i-1}|X_i}) \\ &\quad - \sum_{i=2}^{s} h^0(X_i, \mathcal{H}om(\mathcal{I}_{X_i}, \mathcal{I}_i))\end{aligned}$$

*and*

$$\begin{aligned}\dim_C \operatorname{GradAlg}^H(\mathbb{P}^{n+c}) &= {}_0hom_R(I_{X_s}, B_s) + \sum_{i=2}^{s} {}_0hom_R(I_i, B_{i-1}) \\ &\quad - \sum_{i=2}^{s} h^0(X_i, \mathcal{H}om(\mathcal{I}_{X_i}, \mathcal{I}_i)).\end{aligned}$$

Note that, in the Theorem above, we let ${}_0hom_R(-,-) = \dim {}_0\operatorname{Hom}_R(-,-)$ and we have ${}_0hom_R(I_{X_s}, B_s) = h^0(X_s, \mathcal{N}_{X_s})$ (if $s > 1$) and ${}_0hom_R(I_i, B_{i-1}) = h^0(X_{i-1}, \mathcal{N}_{X_{i-1}|X_i})$ for $i > 2$ because $B_i = \Gamma_*(X_i, \mathcal{O}_{X_i})$ has depth $\geq 2$ at its irrelevant maximal ideal for $i > 2$.

PROOF. In [**38**], (1.6), we saw that the tangent space of "pairs" $(C \subset S)$ of subschemes in $\mathbb{P}^{n+c}$ was given by a cartesian diagram, i.e. essentially given as the diagram (6.1) after having taken global sections. This follows from results of [**46**], Section 2, results which one may adapt to take care of $s$- tuples $(\mathbf{X}.) = (X_1 \subset X_2 \subset \cdots \subset X_s)$ of closed subschemes in $\mathbb{P}^{n+c}$ as well (as done in [**37**]). Indeed if we again let $D(q_1, \ldots, q_s)$ be the Hilbert flag scheme parameterizing $s$-tuples $(\mathbf{X}.)$ of equidimensional, locally Cohen-Macaulay closed subschemes in $\mathbb{P}^{n+c}$, where $q_i$ is the Hilbert polynomial of $X_i$ in $\mathbb{P}^{n+c}$, then the tangent space $A^1(\mathbf{X}.)$ of $D(q_1, \ldots, q_s)$ at $(\mathbf{X}.)$ can be described as follows. Recalling $\mathcal{N}_{X_i} = \mathcal{H}om_{\mathcal{O}_P}(\mathcal{I}_{X_i}, \mathcal{O}_{X_i})$, we consider the diagram

$$\begin{array}{ccccccc} H^0(X_1, \mathcal{N}_{X_1}) & & H^0(X_2, \mathcal{N}_{X_2}) & \ldots & H^0(X_{s-1}, \mathcal{N}_{X_{s-1}}) & & H^0(X_s, \mathcal{N}_{X_s}) \\ & \searrow \quad \swarrow & & \searrow \quad \swarrow & & \searrow \quad \swarrow & \\ & H^0(X_1, \mathcal{H}om_{\mathcal{O}_P}(\mathcal{I}_{X_2}, \mathcal{O}_{X_1})) & & \ldots & & H^0(X_{s-1}, \mathcal{H}om_{\mathcal{O}_P}(\mathcal{I}_{X_s}, \mathcal{O}_{X_{s-1}})). & \end{array}$$

Then $A^1(\mathbf{X}.)$ is the set of elements $(\alpha_1, \alpha_2, \ldots, \alpha_s) \in \sum H^0(X_i, \mathcal{N}_{X_i})$ for which each pair $(\alpha_{i-1}, \alpha_i)$ maps to the same element in $H^0(X_{i-1}, \mathcal{H}om(\mathcal{I}_{X_i}, \mathcal{O}_{X_{i-1}}))$. Since the cokernel of $H^0(\mathcal{N}_{X_i}) \to H^0(X_{i-1}, \mathcal{H}om(\mathcal{I}_{X_i}, \mathcal{O}_{X_{i-1}}))$ is the kernel of $\operatorname{Ext}^1(\mathcal{I}_{X_i}, \mathcal{I}_i) \to \operatorname{Ext}^1(\mathcal{I}_{X_i}, \mathcal{O}_{X_i})$, the diagram above implies the surjectivity of the tangent map $A^1(\mathbf{X}.) \to H^0(X_1, \mathcal{N}_{X_1}) = H^0(\mathcal{N}_C)$ of the first projection

$pr_1 : D(q_1, \ldots, q_s) \to \mathrm{Hilb}^{q_1}(\mathbb{P}^{n+c})$ (which maps $(X_1' \subset X_2' \subset \cdots \subset X_s')$ onto $X_1'$) provided $\mathrm{Ext}^1(\mathcal{I}_{X_i}, \mathcal{I}_i) \to \mathrm{Ext}^1(\mathcal{I}_{X_i}, \mathcal{O}_{X_i})$ is injective for any $i$ (see **[72]**, Proposition 0.4 for a related result). Using Proposition 10.12 and **[26]** (17.11.1), it follows that $pr_1$ is smooth at $(X_1 \subset X_2 \subset \cdots \subset X_s)$, i.e. that $C = X_1$ is unobstructed. Since the whole set-up of this proof works for the graded object $R \to B_s \to \cdots \to B_2 \to A$, provided we replace each $\mathrm{Ext}^i(-,-)$ by ${}_0\mathrm{Ext}^i(-,-)$, we get the $H$-unobstructedness as well.

To prove the first dimension formula, it suffices by the unobstructedness of C to prove that

$$h^0(X_{i-1}, \mathcal{N}_{X_{i-1}}) = h^0(X_{i-1}, \mathcal{N}_{X_{i-1}|X_i}) + h^0(X_i, \mathcal{N}_{X_i}) - h^0(X_i, \mathcal{H}om(\mathcal{I}_{X_i}, \mathcal{I}_i))$$

for $i = 2, 3, \ldots, s$. To show this equality, we essentially need to see that all skew vertical maps of the big diagram above are surjective and to locate their kernels. Since the maps of the big diagram fit into the exact sequences

$$0 \to H^0(X_{i-1}, \mathcal{N}_{X_{i-1}|X_i}) \to H^0(X_{i-1}, \mathcal{N}_{X_{i-1}}) \to H^0(X_{i-1}, \mathcal{H}om(\mathcal{I}_{X_i}, \mathcal{O}_{X_{i-1}}))$$

$$0 \to H^0(X_i, \mathcal{H}om(\mathcal{I}_{X_i}, \mathcal{I}_i)) \to H^0(X_i, \mathcal{N}_{X_i}) \to H^0(X_{i-1}, \mathcal{H}om(\mathcal{I}_{X_i}, \mathcal{O}_{X_{i-1}})) \to 0$$

where the lower one is right exact by assumption, we only need to prove the surjectivity of the right map of the first sequence. To see it we consider the Hilbert flag scheme $D(q_{i-1}, q_i)$ and the second projection $pr_2 : D(q_{i-1}, q_i) \to \mathrm{Hilb}^{q_i}(\mathbb{P}^{n+c})$ given by $(X_{i-1}' \subset X_i') \mapsto (X_i' \subset \mathbb{P}^{n+c})$ on closed points. Its tangent map

$$T_{pr_2} : A^1(X_{i-1}, X_i) \to H^0(X_i, \mathcal{N}_{X_i})$$

at $(X_{i-1} \subset X_i)$ is determined by the cartesian diagram

$$\begin{array}{ccc} A^1(X_{i-1}, X_i) & \xrightarrow{T_{pr_2}} & H^0(X_i, \mathcal{N}_{X_i}) \\ \downarrow & \circ & \downarrow \\ H^0(X_{i-1}, \mathcal{N}_{X_{i-1}}) & \longrightarrow & H^0(X_{i-1}, \mathcal{H}om(\mathcal{I}_{X_i}, \mathcal{O}_{X_{i-1}})) \end{array} \tag{10.16}$$

Since $H^0(X_i, \mathcal{N}_{X_i}) \to H^0(X_{i-1}, \mathcal{H}om(\mathcal{I}_{X_i}, \mathcal{O}_{X_{i-1}}))$ is surjective, it suffices to prove that $T_{pr_2}$ is surjective. Now, let $\lambda \in H^0(X_i, \mathcal{N}_{X_i})$ correspond to an arbitrary deformation $X_{i,T} \subset \mathbb{P}_T^{n+c}$ of $X_i \subset \mathbb{P}^{n+c}$ to the dual numbers $T = k[\epsilon]$. If $i < s$ we know by the first part of this proof that the first projection of $D(q_i, q_{i+1}, \ldots, q_s) \to \mathrm{Hilb}^{q_i}(\mathbb{P}^{n+c})$ is smooth at $(X_i, X_{i+1}, \ldots, X_s)$, i.e. there exists a flag

$$(X_{i,T}, X_{i+1,T}, \ldots, X_{s,T})$$

of $T$-flat schemes reducing to $(X_i, X_{i+1}, \ldots, X_s)$ by putting $\epsilon = 0$. By the proof of Proposition 10.12 the flag $(X_{i,T}, X_{i+1,T}, \ldots, X_{s,T})$ is equipped with the exact sequence

$$0 \to B_{j+1,T}(-f_{j+1}) \to M_{j+1,T} \to M_{j,T} \to 0, \text{ for } i < j < s,$$

of $T$-flat modules such that $\mathcal{M}_{j,T}^*(-f_j)$ is the sheaf ideal $\mathcal{I}_{j,T}$ defining $X_{j-1,T} \subset X_{j,T}$. (If $j = s$, then $\mathcal{M}_{s,T} = \mathcal{I}_{s,T}^*(-f_s)$ and $M_{s,T}$ is the canonical module). As in the final part of the proof of Proposition 10.12 the sections $B_{i+1}(-f_{i+1}) \to M_{i+1}$ (for $i + 1 \leq s$) lift to sections over $T$ and if we let $M_{i,T} = \mathrm{coker}(B_{i+1,T}(-f_{i+1}) \to M_{i+1,T})$, then the twisted dual $\mathcal{M}_{i,T}^*(f_i)$ is a sheaf ideal of $\mathcal{O}_{X_{i,T}}$ which defines a flat deformation $X_{i-1,T} \subset X_{i,T}$ of $X_{i-1} \subset X_i$ to $T = k[\epsilon]$. This shows the surjectivity of $T_{pr_2}$, and hence the first dimension formula.

Moreover the proof of the dimension formula above works for $\mathrm{GradAlg}^H(\mathbb{P}^{n+c})$ at the graded object $R \to B_s \to \cdots \to B_2 \to A$, provided we replace the cohomology groups of the sheaves involved by their corresponding (cohomology) group of graded modules. The proof is therefore complete. □

REMARK 10.14. By the proof above, we see that the surjectivity of $T_{pr_2}$ of (10.16) holds without any injectivity assumptions on the $\mathrm{Ext}^1$-groups if $i = s$. In particular, if $s = 2$ and $(\mathbf{X}.) = (C \subset S \subset \mathbb{P}^{n+3}) \in D(p,q)$, then $T_{pr_2}$ is surjective. Since $(\mathbf{X}.)$ is unobstructed by Proposition 10.12, we can therefore find the dimension $\dim_{(\mathbf{X}.)} D(p,q) = \dim A^1(\mathbf{X}.)$ of $D(p,q)$ at $(\mathbf{X}.)$. Indeed, if $C \subset \mathbb{P}^{n+3}$ is a good determinantal scheme of dimension $n \geq 1$ and $(\mathbf{X}.)$ a flag as in (10.4) and (10.12), we get

$$\dim_{(\mathbf{X}.)} D(p,q) = \dim A^1(\mathbf{X}.) = h^0(S, \mathcal{N}_S) + h^0(C, \mathcal{N}_{C|S}).$$

Hence if $W$ is an irreducible, non-embedded component of $D(p,q)$ containing $(\mathbf{X}.)$, then

$$\dim pr_1(W) = h^0(S, \mathcal{N}_S) + h^0(C, \mathcal{N}_{C|S}) - h^0(S, \mathcal{H}om(\mathcal{I}_S, \mathcal{I}_{C|S}))$$

because the fiber of the composition $W \hookrightarrow D(p,q) \to \mathrm{Hilb}^p(\mathbb{P}^{n+3})$ is $\mathrm{Hom}(\mathcal{I}_S, \mathcal{I}_{C|S})$ at $(\mathbf{X}.)$ (cf. the first part of the proof of Proposition 9.14). This applies to $W(\underline{b};\underline{a})$ because the closures of $W(\underline{b};\underline{a})$ and $pr_1(W)$ in $\mathrm{Hilb}^p(\mathbb{P}^{n+3})$ are the same. Indeed, to prove this equality of closures, let $C \subset S$ be good determinantal schemes, defined by $[M, N]$ and $N$ respectively, where $S$ is given by deleting say the *first* column, $M$, and let $W_1(\underline{b};\underline{a})$ be the locus of good determinantal "pairs" $C' \subset S'$ of $D(p,q)$ defined by matrices $[M', N']$ and $N'$ respectively whose components have the same degree as the matrices of $C$ and $S$. Then we claim that $\overline{W_1(\underline{b};\underline{a})}$ is an irreducible component of $D(p,q)$ (contrary to what we saw for $W(\underline{b};\underline{a})$ in $\mathrm{Hilb}^p(\mathbb{P}^{n+3})$; cf. Example 10.5 (4)). Indeed, applying the notation and essentially the same arguments as we used when we defined $\mathbb{Y}^0 \subset \mathbb{V}$ on page 96 and the map $\mathbb{Y}^0 \to W(\underline{b};\underline{a})$ on page 97, we see that there exists a non-empty open subset $\mathbb{Y}^1$ of $\mathbb{V}$ and a morphism $\mathbb{Y}^1 \to D(p,q)$ inducing a dominating map $\mathbb{Y}^1 \to W_1(\underline{b};\underline{a})$. Moreover, $\mathbb{Y}^1 \to D(p,q)$ is smooth at $[M, N]$ because it is surjective on tangent spaces. This surjectivity follows from the proof of the claim to which (10.13) belongs. Indeed, a deformation of $(C_T \subset S_T \subset \mathbb{P}^{n+3}{}_T)$ of $(\mathbf{X}.)$ in $D(p,q)$ leads to an exact sequence (10.14) for $i+1 = s$, and as in (3.1), to a matrix $[M_T, N_T]$ defining $C_T$ and $S_T$. Since a smooth map takes *generic* points onto generic points, we see that $\overline{W_1(\underline{b};\underline{a})}$ is an irreducible component of $D(p,q)$. Finally, since the restriction of $pr_1$ to $W_1(\underline{b};\underline{a})$, namely $W_1(\underline{b};\underline{a}) \to W(\underline{b};\underline{a})$, is dominating, we get $\overline{W(\underline{b};\underline{a})} = \overline{pr_1(W)}$, as indicated in Remark 10.6.

As a consequence of Theorem 10.13 we will prove Proposition 10.7 under weaker conditions than previously. Indeed we are able to replace the Cartier assumption on $C \subset S$ by Cartier in codimension 3 in $S$, an important generalization which removes the application of Proposition 10.7 to small dimensional cases only (i.e. to $n \leq 2$) because a general ACM scheme $S$ of codimension 2 in $\mathbb{P}^{n+3}$ is smooth in codimension 3. Moreover we will prove a similar result for the zero-dimensional codimension 3 case. Finally we illustrate that Theorem 10.13 takes care of determinantal subschemes of codimension at least 4 by considering an example. Note that the terms $h^0(\mathcal{N}_{X_{i-1}|X_i})$ of the dimension formula of Theorem 10.13 are quite easy to compute using (10.15) and (10.5) and $h^0(\mathcal{N}_{X_s})$ is of course well known while, unfortunately, the term $h^0(\mathcal{H}om(\mathcal{I}_{X_i}, \mathcal{I}_i))$ and the vanishing of $\mathrm{Ext}^1(\mathcal{I}_{X_i}, \mathcal{I}_i)$ have been harder to make explicit. For these groups we have at least the vanishing result of Proposition 9.7. In codimension 3, however, we have the following quite complete result

COROLLARY 10.15. *Let $C \subset S$ be good determinantal schemes in $\mathbb{P}^{n+3}$ as in (10.12), where $C$ is given by the maximal minors of a $t \times (t+2)$ matrix whose $ij$-th entry is homogeneous of degree $a_j - b_i, 0 \le j \le t+1, 1 \le i \le t$, and $S$ is correspondingly given by deleting the $(k+1)$-st column. Let*

$$l = \sum_{j=0}^{t+1} a_j - \sum_{i=1}^{t} b_i,$$

*$a_r = \max\{a_j\}$ and $n = \dim C \ge 1$ and suppose $C \subset S$ is a local complete intersection (i.e. Cartier) outside some $Z$ of codimension $\ge 3$ in $C$. Then*

(i) $\dim W(\underline{b}; \underline{a}) = \lambda + \kappa$ *with $\lambda$ and $\kappa$ as in Proposition 10.7.*
(ii) *If $\dim C > 1$, or if $\dim C = 1$ and $H^1(\mathcal{I}_S/\mathcal{I}_S^2(l - 2a_k)) = 0$, then $C$ is unobstructed and*

$$\dim_C \mathrm{Hilb}^p(\mathbb{P}^{n+3}) = \lambda + \kappa.$$

PROOF. Since one of the dimension formulas really requires $C$ to be unobstructed, we consider the unobstructedness of (ii) first. We can suppose $\dim C \ge 3$ because the case $\dim C \le 2$ is taken care of in Proposition 10.7. To prove that $C$ is unobstructed, it suffices by Theorem 10.13 to show the injectivity of

$$\mathrm{Ext}^1(\mathcal{I}_S, \mathcal{I}_{C|S}) \to \mathrm{Ext}^1(\mathcal{I}_S, \mathcal{O}_S),$$

or equivalently, the surjectivity of $\mathrm{Hom}(\mathcal{I}_S, \mathcal{O}_S) \to \mathrm{Hom}(\mathcal{I}_S, \mathcal{O}_C)$. Since $\mathrm{depth}_{I_Z} B \ge 2$ and $\mathrm{depth}_{I_Z} A \ge 2$, we have two isomorphisms fitting into

$$(10.17)\qquad \begin{array}{ccc} \mathrm{Hom}_{\mathcal{O}_P}(\mathcal{I}_S, \mathcal{O}_S) & \cong & H^0(S - Z, \mathcal{H}om_{\mathcal{O}_P}(\mathcal{I}_S, \mathcal{O}_S)) \\ \downarrow & \circ & \downarrow \\ \mathrm{Hom}_{\mathcal{O}_P}(\mathcal{I}_S, \mathcal{O}_C) & \cong & H^0(S - Z, \mathcal{H}om_{\mathcal{O}_P}(\mathcal{I}_S, \mathcal{O}_C)) \end{array}$$

and so it suffices to show that $H^1(S - Z, \mathcal{H}om(\mathcal{I}_S, \mathcal{I}_{C|S})) = 0$ because

$$\mathcal{H}om(\mathcal{I}_S, \mathcal{O}_S) \to \mathcal{H}om(\mathcal{I}_S, \mathcal{O}_C)$$

is surjective on $S - Z$. Indeed on $S - Z$ we have

$$\mathcal{E}xt^1_{\mathcal{O}_S}(\mathcal{I}_S/\mathcal{I}_S^2, \mathcal{I}_{C|S}) \cong \mathcal{E}xt^1_{\mathcal{O}_S}(\mathcal{I}_S/\mathcal{I}_S^2, \mathcal{O}_S) \otimes \mathcal{I}_{C|S} = 0$$

by Proposition 6.17 because

$$\mathcal{E}xt^1(\mathcal{I}_S/\mathcal{I}_S^2, \mathcal{O}_S) \hookrightarrow H^2(R, B, B)^\sim.$$

As in the proof of Proposition 10.7, we get

$$H^1(S - Z, \mathcal{H}om(\mathcal{I}_S, \mathcal{I}_{C|S})) \cong H^1(S - Z, \mathcal{I}_S/\mathcal{I}_S^2(n + 4 - f)) = 0.$$

Moreover $R \to B$ is generically a complete intersection, and it follows from Remark 10.9 that there is an exact sequence

$$0 \to H_1 \to K_1 \otimes_R B \to I_S/I_S^2 \to 0$$

where $\mathbf{K}.$ is the Koszul complex of a set of minimal generators of $I_S$ and $H_1$ its first homology. Moreover [**3**], Theorem 2.1, implies that $H_1$ is a maximal Cohen-Macaulay module (of projective dimension 2 over $R$) and that

$$\mathrm{depth}_{I_Z} H_1 = \mathrm{depth}_{I_Z} B$$

which we know is at least 4 by assumption. Hence we get the required vanishing from the exact sequence

$$H^1(S,\mathcal{I}_S/\mathcal{I}_S^2(n+4-f)) \to H^1(S-Z,\mathcal{I}_S/\mathcal{I}_S^2(n+4-f)) \to H_Z^2(\mathcal{I}_S/\mathcal{I}_S^2(n+4-f))$$

and

$$H_Z^2(\mathcal{I}_S/\mathcal{I}_S^2(n+4-f)) \cong H_Z^3(H_1^{\sim}(n+4-f)) = 0$$

and from

$$H^1(S,\mathcal{I}_S/\mathcal{I}_S^2(n+4-f)) \cong H^2(S,H_1^{\sim}(n+4-f)) = 0.$$

$C$ is therefore unobstructed by Theorem 10.13.

To see that the dimension formulas of (i) and (ii) hold, we claim that the dimension of $\mathrm{Hilb}^p(\mathbb{P}^{n+3})$ (resp. of $W(\underline{b};\underline{a})$) given by Theorem 10.13 (resp. by Remark 10.14) coincides with the right hand side of (10.6). Indeed by (10.15) we have

$$H^0(\mathcal{N}_{C|S}) \cong H^0(S,\mathcal{M}_s(f))/k^* \cong H^0(\omega_S(f))/k^*,$$

$$(10.18)\qquad \begin{aligned} \mathrm{Hom}(\mathcal{I}_S,\mathcal{I}_{C|S}) &\cong H^0(S-Z,\mathcal{H}om(\mathcal{I}_S,\mathcal{I}_{C|S})) \\ &\cong H^0(S-Z,\mathcal{I}_S/\mathcal{I}_S^2(n+4-f)) \\ &\cong H^0(S,\mathcal{I}_S/\mathcal{I}_S^2(n+4-f)) \\ &\cong (I_S/I_S^2)_{n+4-f}, \end{aligned}$$

because the modules involved have sufficient depth at $I_Z$. The formulas of $\dim_C \mathrm{Hilb}^p(\mathbb{P}^{n+3})$ (resp. of $W(b;a)$) of Theorem 10.13 and Remark 10.14 are therefore equal to

$$h^0(\omega_S(f)) - 1 + h^0(\mathcal{N}_S) - \dim(I_S/I_S^2)_{n+4-f}$$

and the claim is proved. Since the dimension calculations in the final part of the proof of Proposition 10.7 hold without requiring $C \subset S$ to be Cartier, we get the corollary. □

EXAMPLE 10.16. Let $C$ be a good determinantal curve in $\mathbb{P}^5$ given by the maximal minors of a $t \times (t+3)$ matrix $[L, L', M]$ where $M$ (resp. $L'$) is a column of entries of degree $m$ (resp. degree 1) and $L$ (resp. $[L, L']$) is a matrix of linear entries whose maximal minors define a subscheme $X$ (resp. $S$), i.e. we have a flag $(\mathbf{X}.) = (C \subset S \subset X \subset \mathbb{P}^5)$, $X_1 = C$, $X_2 = S$ etc. which we suppose satisfy (10.12). If $S \subset X$ is a Cartier divisor in codimension 3 in $X$, then the map $\mathrm{Ext}^1(\mathcal{I}_X,\mathcal{I}_{S|X}) \to \mathrm{Ext}^1(\mathcal{I}_X,\mathcal{O}_X)$ of Theorem 10.13 is injective by Corollary 10.15. To get $\mathrm{Ext}^1(\mathcal{I}_S,\mathcal{I}_{C|S}) = 0$, it suffices by Proposition 9.7 to verify

$$s(C|S) \geq 5 + a$$

where $a = max\{v | H^2(\mathcal{O}_C(v)) \neq 0\} + 1$. Using the computations of Example 7.5 we get $a = t-3$. Since the Eagon-Northcott complex associated to the determinantal curve $C$ lead to $s(C|S) = m+t-1$, we get the vanishing of $\mathrm{Ext}^1(\mathcal{I}_S,\mathcal{I}_{C|S})$ and the unobstructedness of $C$ for $m \geq 3$. Note that by Proposition 9.7 we also have $\mathrm{Hom}(\mathcal{I}_S,\mathcal{I}_{C|S}) = 0$ which simplifies the dimension formula of Theorem 10.13 to

$$\dim_C \mathrm{Hilb}^p(\mathbb{P}^5) = h^0(S,\mathcal{N}_S) + h^0(C,\mathcal{N}_{C|S})$$

because

$$h^0(S,\mathcal{N}_S) = h^0(X,\mathcal{N}_X) + h^0(\mathcal{N}_{S|X}) - h^0(X,\mathcal{H}om(\mathcal{I}_X,\mathcal{I}_{S|X})).$$

Moreover by Example 10.10 we have

$$\dim_S \operatorname{Hilb}^p(\mathbb{P}^5) = h^0(\mathcal{N}_S) = 4t^2 + 8t - 3,$$

and it remains to compute $h^0(\mathcal{N}_{C|S})$. Using (10.15) for $i = 2$ and (10.5) for $i = 3$, we remark that a careful deletion of columns of our matrix leads, via an argument similar to that used with the diagram (3.1), to

$$f_2 = 5 + m - t \quad \text{and} \quad f_3 = 6 - t.$$

Then we get

$$h^0(\mathcal{N}_{C|S}) = h^0(\mathcal{M}_S(f_2)) - 1 = h^0(\omega_X(f_2)) - h^0(\mathcal{O}_X(f_2 - f_3)) - 1.$$

We have

$$h^0(\mathcal{O}_X(f_2 - f_3)) = h^0(\mathcal{O}_X(m-1)) = \binom{m+4}{5} - h^0(\mathcal{I}_X(m-1))$$

where we, if necessary, can calculate $h^0(\mathcal{I}_X(m-1))$ by a sequence similar to (10.7). By (10.8),

$$h^0(\omega_X(f_2)) = \binom{m+5}{5}t - \binom{m+4}{5}(t+1) + \binom{m-t+4}{5},$$

and we get

$$\begin{aligned} \dim_C \operatorname{Hilb}^p(\mathbb{P}^5) &= h^0(\mathcal{N}_S) + h^0(\mathcal{N}_{C|S}) \\ &= 4t^2 + 8t - 4 + \binom{m+5}{5}t - \binom{m+4}{5}(t+2) \\ &\quad + \binom{m-t+4}{5} + h^0(\mathcal{I}_X(m-1)). \end{aligned}$$

If $3 \le m \le t$, then $h^0(\mathcal{I}_X(m-1)) = 0$ and $\binom{m-t+4}{5} = 0$, in which case one may check that the dimension above is precisely equal to the formula $\lambda$ of Proposition 10.7 (where now all $i, j$ means $1 \le i \le t$ and $0 \le j \le t+2$); cf. Example 10.5.

Our final corollary is mainly concerned with $H$-unobstructedness in the zero-dimensional codimension 3 case. It is based on Proposition 9.7 and Remark 9.8 (i) which implies the vanishing of $\operatorname{Ext}^1(\mathcal{I}_S, \mathcal{I}_{C|S})$ (and of ${}_0\operatorname{Ext}^1_R(I_S, I_{C|S})$ !) provided the middle term of the complex (7.2) vanishes. Since Remark 9.8 (i) does not require $C$ to be Cartier on $S$, it follows that we can also generalize the 1 dimensional case of Corollary 10.11 (allowing us to weaken the Cartier assumption) as well.

COROLLARY 10.17. *Let $C \subset S$ be good determinantal schemes in $\mathbb{P}^{n+3}$ as in (10.12) where $C$ is given as the maximal minors of a $t \times (t+2)$ matrix whose $ij$-th entry is homogeneous of degree $a_j - b_i, 0 \le j \le t+1, 1 \le i \le t$, and $S$ is correspondingly given by deleting the first column. Let*

$$b_1 \ge \cdots \ge b_t, \quad a_0 \ge a_1 \ge \cdots \ge a_t \ge a_{t+1} \quad \text{and} \quad l = \sum_{j=0}^{t+1} a_j - \sum_{i=1}^{t} b_i,$$

*and suppose $a_0 > a_1 + a_2 - b_t$. If $n = \dim C = 0$ (resp. $n = 1$), then*

$${}_0\operatorname{Ext}^1_R(I_S, I_{C|S}) = 0 \quad (\textit{resp.} \quad \operatorname{Ext}^1(\mathcal{I}_S, \mathcal{I}_{C|S}) = 0),$$

*i.e. $C$ is $H$-unobstructed (resp. unobstructed), and we have*

$$\dim_C \operatorname{GradAlg}^H(\mathbb{P}^3) = \lambda + \kappa,$$

$$\dim_C \operatorname{Hilb}^p(\mathbb{P}^3) - \dim_C \operatorname{GradAlg}^H(\mathbb{P}^3) = h^1(\mathcal{I}_S/\mathcal{I}_S^2(l-2a_0)) - h^1(\mathcal{N}_S)$$

$$(\text{resp.} \quad \dim_C \operatorname{Hilb}^p(\mathbb{P}^4) = \lambda + \kappa)$$

*where $\lambda$ and $\kappa$ are given in Proposition 10.7.*

PROOF. To prove that $C$ is $H$-unobstructed (resp. unobstructed), it suffices by Theorem 10.13 to show that ${}_0\operatorname{Ext}^1_R(I_S, I_{C|S}) = 0$ (resp. $\operatorname{Ext}^1(\mathcal{I}_S, \mathcal{I}_{C|S}) = 0$). Since the maximal minors of the $t \times (t+2)$ matrix above which are not minimal generators of $I_S$, have degree $l - a_j - a_k$, $1 \le j < k \le t+1$, we get $(I_{C|S})_{l-a_0-b_i} = 0$ for any $i$ by the assumption $a_0 > a_1 + a_2 - b_t$ (cf. proof of Corollary 10.11). The mentioned Ext-groups therefore vanish by (7.3) and (7.4).

To get the dimension formula, it suffices by the proof of Proposition 10.7 to see that (10.6) and the dimension formula of Theorem 10.13 coincide. As in (10.18), we get

$${}_0\operatorname{Hom}_R(I_{C|S}, A) \cong (K_B)_f/k^*, (\textit{resp. } H^0(\mathcal{N}_{C|S}) \cong H^0(\omega_S(f))/k^*)$$

while

$${}_0\operatorname{Hom}_R(I_S, I_{C|S}) = (I_S/I_S^2)_{n+4-f}$$

(respectively

$$\operatorname{Hom}(\mathcal{I}_S, \mathcal{I}_{C|S}) = (I_S/I_S^2)_{n+4-f})$$

follows from the fact that all these groups vanish. (cf. Proposition 9.7 for the Hom-groups, and for the other we observe $(I_S)_{l-2a_0} = 0$ because $a_2 > b_t$ by assumption).

Finally the codimension formula in the zero dimensional case (in which case $C \subset S$ is Cartier) follows from Theorem 9.16 (i.e. from the last part of its proof) because $H^1(\mathcal{I}_S/\mathcal{I}_S^2(l-2a_0)) \cong H^1(\mathcal{N}_S \otimes \mathcal{I}_{C|S})$. □

EXAMPLE 10.18. Let $C$ be a good determinantal zero dimensional subscheme of $\mathbb{P}^3$ given by the maximal minors of a $t \times (t+2)$ matrix $[M, L]$ where $M$ is a column of entries of degree $m$ and $L$ is a matrix of linear entries whose minors define a subscheme $S$ as in (10.12). If $m > 2$, then $C$ is $H$-unobstructed and $\dim_C \operatorname{GradAlg}^H(\mathbb{P}^3)$ equals the formula of Example 10.10 with $n = 0$. (Of course Corollary 10.17 also shows that the conclusions of Example 10.10 hold if $\dim C = 1$, only requiring $C \subset S$ to be generically Cartier and $S$ to satisfy $G_1$).

# Bibliography


[1] R. Apéry, *Sur certain caractères numériques d'un idéal sans composant impropre*, C.R.A.S. 220 (1945), 234–236.

[2] R. Apéry, *Sur les courbes de première espèce de l'espace à trois dimensions*, C.R.A.S. 220 (1945), 271–272.

[3] L. Avramov and J. Herzog, *The Koszul Algebra of a Codimension 2 Embedding*, Math. Z. **175** (1980), 249–260.

[4] E. Ballico, G. Bolondi and J. Migliore, *The Lazarsfeld-Rao Problem for Liaison Classes of Two-Codimensional Subschemes of* $\mathbb{P}^n$, Amer. J. of Math. **113** (1991), 117–128.

[5] D. Bayer and M. Stillman, *Macaulay*, a computer system for computing in Commutative Algebra and Algebraic Geometry.

[6] M. Boij, *Gorenstein Artin algebras and points in projective space*, Bull. London Math. Soc. **31** (1999), 11–16.

[7] G. Bolondi and J. Migliore, *The Structure of an Even Liaison Class*, Trans. Amer. Math. Soc. **316** (1989), 1–37.

[8] G. Bolondi and J. Migliore, *The Lazarsfeld-Rao property on an arithmetically Gorenstein variety*, Manuscripta Math. 78 (1993), 347–368.

[9] W. Bruns, *The Eisenbud-Evans generalized principal ideal theorem and determinantal ideals*, Proc. Amer. Math. Soc. **83** (1981), 19–24.

[10] W. Bruns, J. Herzog, *"Cohen-Macaulay rings"*, Cambridge Studies in Advanced Mathematics **39**, Cambridge University Press, 1993.

[11] W. Bruns and U. Vetter, "Determinantal Rings," Lecture Notes in Math. **1327**, Springer-Verlag, 1988.

[12] D. Buchsbaum and D. Eisenbud, *What annihilates a module?*, J. Algebra **47** (1977), 231–243.

[13] R.O. Buchweitz, "Contributions a la theorie des singularites," Thesis l'Université Paris VII (1981).

[14] R.O. Buchweitz and B. Ulrich, *Homological properties which are invariant under linkage*, Preprint 1983 .

[15] M. Casanellas and R.M. Miró-Roig, *Gorenstein Liaison of curves in* $\mathbb{P}^4$, J. Algebra **230** (2000), 656–664.

[16] M. Casanellas and R.M. Miró-Roig, *Gorenstein liaison of divisors on standard determinantal schemes and on rational normal scrolls*, to appear in J. Pure Appl. Algebra.

[17] E. Davis, A.V. Geramita, F. Orecchia, *Gorenstein Algebras and the Cayley-Bacharach Theorem*, Proc. Amer. Math. Soc. **93** (1985), 593–597.

[18] T. de Jong and D. van Straten, *Deformations of normalization of hypersurfaces*, Math. Ann. *288* (1990), 527–547.

[19] P. Dubreil, *Quelques propriétés des variétés algébriques*, Actualités Scientifiques et Industrielles **210**, Paris: Hermann, 1935.

[20] L. Ein, *Hilbert scheme of smooth space curves*, Ann. Scient. Ec. Norm. sup. **19** (1986), 469–478.

[21] D. Eisenbud, "Commutative Algebra with a view toward Algebraic Geometry," Graduate Texts in Mathematics **150**, Springer-Verlag, 1995.

[22] G. Ellingsrud, *Sur le schema de Hilbert des variétés de codimension 2 dans* $\mathbb{P}^e$ *a cone de Cohen-Macaulay*, Ann. Scient. Ec. Norm. Sup. **8** (1975), 423–432.

[23] E.G. Evans and P.A. Griffith, *Local cohomology modules for normal domains*, J. London Math. Soc., **22** (1979), 277–284.

[24] F. Gaeta, *Nuove ricerche sulle curve sghembe algebriche di residuale finito e sui gruppi di punti del piano*, Ann. di Mat. Pura et Appl., ser.4, **31** (1950), 1–64.

[25] A.V. Geramita and J. Migliore, *A Generalized Liaison Addition*, J. Algebra **163** (1994), 139–164.

[26] A. Grothendieck, "Elements de Geometrie Algebrique IV," Springer-Verlag, 1971.

[27] A. Grothendieck, *Technique de descente et theoreme d'existence en geometrie algebrique*, in Sem. Bourbaki, Vol 1960/1961, exp. 221.

[28] A. Grothendieck, "Cohomologie locale des faisceaux coherents et Theoremes de Lefschetz locaux et globaux," North Holland, Amsterdam, 1968.

[29] R. Hartshorne, "Local Cohomology," Lecture Notes in Math. **41**, Springer-Verlag, 1967.

[30] R. Hartshorne, *Generalized Divisors on Gorenstein Schemes*, $K$-Theory **8** (1994), 287–339.

[31] R. Hartshorne, *Algebraic Geometry*, Springer-Verlag, GTM **52**.

[32] J. Herzog, *Deformationen von Cohen-Macaulay Algebren*, J. reine angew. Math. **318** (1980), 83–105.

[33] J. Herzog, *Ein Cohen-Macaulay-Kriterium mit Anwendungen auf den Konormalenmodul und den Differentialmodul*, Math. Zeit. **163** (1978), 149–162.

[34] H.J. Hoppe, *Generischer Spaltungstyp und zweite ChernKlasse stabiler Vektorraumbundel vom Rang 4 auf* $\mathbb{P}^4$, Math. Zeit. **187** (1984), 345-360.

[35] C. Huneke, *Numerical invariants of liaison classes*, Invent. Math. **75** (1984), 301–325.

[36] C. Huneke, B. Ulrich, *The structure of linkage*, Ann. Math. **126** (1987), 277–334.

[37] J.O. Kleppe, *The Hilbert-flag scheme, its properties and its connection with the Hilbert scheme. Applications to curves in 3-space*, Preprint (part of thesis), March 1981, Univ. of Oslo.

[38] J.O. Kleppe, *Liaison of families of subschemes in* $\mathbb{P}^n$, In: Proc. Trento 1988. Lecture Notes in Math. **1389**, Springer-Verlag, 1989.

[39] J.O. Kleppe, *Concerning the existence of nice components in the Hilbert scheme curves in* $\mathbb{P}^n$ *for* $n = 4$ *and 5*, J. reine angew. Math. **475** (1996), 77–102.

[40] J.O. Kleppe, *Deformations of graded algebras*, Math. Scand. **45** (1979).

[41] J.O. Kleppe, *The smoothness and the dimension of* $PGor(H)$ *and of other strata of the punctual Hilbert scheme*, J. Algebra **200** (1998), 606–628.

[42] J.O. Kleppe, *On the existence of nice components in the Hilbert scheme* $H(d, g)$ *of smooth connected space curves*, Boll. U.M.I. **(7) 8-B** (1994), 305–326.

[43] J.O. Kleppe, *Halphen gaps and good space curves*, Boll. U.M.I. (8) 1-B (1998), 429–450.

[44] J.O. Kleppe and R.M. Miró-Roig, *The dimension of the Hilbert scheme of Gorenstein codimension 3 subschemes*, J. Pure Appl. Algebra **127** (1998), 73–82.

[45] M. Kreuzer, J. Migliore, U. Nagel and C. Peterson, *Determinantal Schemes and Buchsbaum-Rim Sheaves*, to appear in J. Pure Appl. Algebra.

[46] O.A. Laudal, *A generalized trisecant lemma*, Proc. Tromsø, Conference in algebraic geometry, Lecture Notes in Math. **754**, Springer, 1977, pp. 112–149.

[47] O.A. Laudal, "Formal Moduli of Algebraic Structures," Lecture Notes in Math. **754**, Springer-Verlag, 1978.

[48] R. Lazarsfeld and P. Rao, *Linkage of General Curves of Large Degree*, in "Algebraic Geometry– Open Problems (Ravello, 1982)," Lecture Notes in Mathematics, vol. 997, Springer–Verlag, 1983, pp. 267–289.

[49] M. Martin-Deschamps and D. Perrin, "Sur la Classification des Courbes Gauches," Astérisque 184–185, Soc. Math. de France (1990).

[50] M. Martin-Deschamps and R. Piene, *Arithmetically Cohen-Macaulay curves in* $\mathbb{P}^4$ *of degree 4 and genus 0*, Manuscripta Math. **93** (1997), 391–408.

[51] J. Migliore, *Geometric Invariants for Liaison of Space Curves*, J. Algebra **99** (1986), 548–572.

[52] J. Migliore, "Introduction to Liaison Theory and Deficiency Modules," Birkhäuser, Progress in Mathematics 165 (1998).

[53] J. Migliore and U. Nagel, *Monomial ideals and the Gorenstein liaison class of a complete intersection*, preprint.

[54] J. C. Migliore, U. Nagel, C. Peterson, *Buchsbaum-Rim sheaves and their multiple sections*, J. Algebra **219** (1999), 378–420.

[55] J. Migliore and C. Peterson, *A construction of codimension three arithmetically Gorenstein subschemes of projective space*, Trans. Amer. Math. Soc. **349** (1997), 3803–3821.

[56] U. Nagel, *On Hilbert Function Under Liaison*, Le Matematiche XLVI (1991), 547–558.

[57] U. Nagel, *Even Liaison Classes Generated by Gorenstein Linkage*, J. Algebra **209** (1998), 543–584.

[58] S. Nollet, *Even Linkage Classes*, Trans. Amer. Math. Soc. **348** (1996), no. 3, 1137–1162.

[59] R. Notari and M.L. Spreafico, *A Stratification of Hilbert Scheme by Initial Ideals and Applications*, Manuscripta Math. **101** (2000), 429-448.

[60] C. Okonek, *Reflexive Garben auf* $\mathbb{P}^4$, Math. Ann. **260** (1982), 211-237.

[61] C. Okonek, *Moduli reflexiver Garben und Flächen vom kleinem Grad in* $\mathbb{P}^4$, Math. Zeit. **184** (1983), 549–572.

[62] C. Peskine and L. Szpiro, *Liaison des variétés algébriques. I*, Invent. Math. **26** (1974), 271–302.

[63] P. Rao, *Liaison among Curves in* $\mathbb{P}^3$, Invent. Math. **50** (1979), 205–217.

[64] P. Rao, *Liaison Equivalence Classes*, Math. Ann. **258** (1981), 169–173.

[65] P. Rao, *On Self-Linked Curves*, Duke Math. J. **49(2)** (1982), 251–273.

[66] P. Schenzel, *Notes on Liaison and Duality*, J. Math. Kyoto Univ. **22** (1982), 485–498.

[67] P. Schwartau, *Liaison Addition and Monomial Ideals*, Ph.D. thesis, Brandeis University (1982).

[68] T. Svanes, *Some Criteria for Rigidity of Noetherian Rings*, Math. Z. **144** (1975), 135–145.

[69] B. Ulrich, *Rings of Invariants and Linkage of Determinantal Ideals*, Math. Ann. **274** (1986), 1–17.

[70] W. Vasconcelos, "Arithmetic of Blowup Algebras," London Math. Soc. Lecture Note Series **195**, Cambridge University Press (1994).

[71] C. Walter, *Cohen-Macaulay liaison*, J. reine angew. Math. **444** (1993), 101–114.

[72] C. Walter, *Components of the stack of torsion-free sheaves of rank 2 on ruled surfaces*, Math. Ann. **301** (1995), 699-715.

[73] J. Watanabe, *A note on Gorenstein rings of embedding codimension 3*, Nagoya Math. J. **50** (1973), 227–232.

[74] C. Weibel, "An introduction to homological algebra," Cambridge University Press, Cambridge studies in advanced mathematics 38 (1994).

## Editorial Information

To be published in the *Memoirs*, a paper must be correct, new, nontrivial, and significant. Further, it must be well written and of interest to a substantial number of mathematicians. Piecemeal results, such as an inconclusive step toward an unproved major theorem or a minor variation on a known result, are in general not acceptable for publication. Papers appearing in *Memoirs* are generally longer than those appearing in *Transactions*, which shares the same editorial committee.

As of May 31, 2001, the backlog for this journal was approximately 7 volumes. This estimate is the result of dividing the number of manuscripts for this journal in the Providence office that have not yet gone to the printer on the above date by the average number of monographs per volume over the previous twelve months, reduced by the number of volumes published in four months (the time necessary for preparing a volume for the printer). (There are 6 volumes per year, each containing at least 4 numbers.)

A Consent to Publish and Copyright Agreement is required before a paper will be published in the *Memoirs*. After a paper is accepted for publication, the Providence office will send a Consent to Publish and Copyright Agreement to all authors of the paper. By submitting a paper to the *Memoirs*, authors certify that the results have not been submitted to nor are they under consideration for publication by another journal, conference proceedings, or similar publication.

## Information for Authors

*Memoirs* are printed from camera copy fully prepared by the author. This means that the finished book will look exactly like the copy submitted.

The paper must contain a *descriptive title* and an *abstract* that summarizes the article in language suitable for workers in the general field (algebra, analysis, etc.). The *descriptive title* should be short, but informative; useless or vague phrases such as "some remarks about" or "concerning" should be avoided. The *abstract* should be at least one complete sentence, and at most 300 words. Included with the footnotes to the paper should be the 2000 *Mathematics Subject Classification* representing the primary and secondary subjects of the article. The classifications are accessible from `www.ams.org/msc/`. The list of classifications is also available in print starting with the 1999 annual index of *Mathematical Reviews*. The Mathematics Subject Classification footnote may be followed by a list of *key words and phrases* describing the subject matter of the article and taken from it. Journal abbreviations used in bibliographies are listed in the latest *Mathematical Reviews* annual index. The series abbreviations are also accessible from `www.ams.org/publications/`. To help in preparing and verifying references, the AMS offers MR Lookup, a Reference Tool for Linking, at `www.ams.org/mrlookup/`. When the manuscript is submitted, authors should supply the editor with electronic addresses if available. These will be printed after the postal address at the end of the article.

**Electronically prepared manuscripts.** The AMS encourages electronically prepared manuscripts, with a strong preference for $\mathcal{A}_\mathcal{M}\mathcal{S}$-LaTeX. To this end, the Society has prepared $\mathcal{A}_\mathcal{M}\mathcal{S}$-LaTeX author packages for each AMS publication. Author packages include instructions for preparing electronic manuscripts, the *AMS Author Handbook*, samples, and a style file that generates the particular design specifications of that publication series. Though $\mathcal{A}_\mathcal{M}\mathcal{S}$-LaTeX is the highly preferred format of TeX, author packages are also available in $\mathcal{A}_\mathcal{M}\mathcal{S}$-TeX.

Authors may retrieve an author package from e-MATH starting from `www.ams.org/tex/` or via FTP to `ftp.ams.org` (login as `anonymous`, enter username as password, and type `cd pub/author-info`). The *AMS Author Handbook* and the *Instruction Manual* are available in PDF format following the author packages link from `www.ams.org/tex/`. The author package can be obtained free of charge by sending email to `pub@ams.org` (Internet) or from the Publication Division, American Mathematical Society, P.O. Box 6248, Providence, RI 02940-6248. When requesting an author package, please specify AMS-LaTeX or AMS-TeX, Macintosh or IBM (3.5) format, and the publication in which your paper will appear. Please be sure to include your complete mailing address.

**Sending electronic files.** After acceptance, the source file(s) should be sent to the Providence office (this includes any TeX source file, any graphics files, and the DVI or PostScript file).

Before sending the source file, be sure you have proofread your paper carefully. The files you send must be the EXACT files used to generate the proof copy that was accepted for publication. For all publications, authors are required to send a printed copy of their paper, which exactly matches the copy approved for publication, along with any graphics that will appear in the paper.

TeX files may be submitted by email, FTP, or on diskette. The DVI file(s) and PostScript files should be submitted only by FTP or on diskette unless they are encoded properly to submit through email. (DVI files are binary and PostScript files tend to be very large.)

Electronically prepared manuscripts can be sent via email to `pub-submit@ams.org` (Internet). The subject line of the message should include the publication code to identify it as a Memoir. TeX source files, DVI files, and PostScript files can be transferred over the Internet by FTP to the Internet node `e-math.ams.org` (130.44.1.100).

**Electronic graphics.** Comprehensive instructions on preparing graphics are available at `www.ams.org/jourhtml/graphics.html`. A few of the major requirements are given here.

Submit files for graphics as EPS (Encapsulated PostScript) files. This includes graphics originated via a graphics application as well as scanned photographs or other computer-generated images. If this is not possible, TIFF files are acceptable as long as they can be opened in Adobe Photoshop or Illustrator. No matter what method was used to produce the graphic, it is necessary to provide a paper copy to the AMS.

Authors using graphics packages for the creation of electronic art should also avoid the use of any lines thinner than 0.5 points in width. Many graphics packages allow the user to specify a "hairline" for a very thin line. Hairlines often look acceptable when proofed on a typical laser printer. However, when produced on a high-resolution laser imagesetter, hairlines become nearly invisible and will be lost entirely in the final printing process.

Screens should be set to values between 15% and 85%. Screens which fall outside of this range are too light or too dark to print correctly. Variations of screens within a graphic should be no less than 10%.

**Inquiries.** Any inquiries concerning a paper that has been accepted for publication should be sent directly to the Electronic Prepress Department, American Mathematical Society, P. O. Box 6248, Providence, RI 02940-6248.

## Editors

This journal is designed particularly for long research papers, normally at least 80 pages in length, and groups of cognate papers in pure and applied mathematics. Papers intended for publication in the *Memoirs* should be addressed to one of the following editors. In principle the Memoirs welcomes electronic submissions, and some of the editors, those whose names appear below with an asterisk (*), have indicated that they prefer them. However, editors reserve the right to request hard copies after papers have been submitted electronically. Authors are advised to make preliminary email inquiries to editors about whether they are likely to be able to handle submissions in a particular electronic form.


**Algebra** to CHARLES CURTIS, Department of Mathematics, University of Oregon, Eugene, OR 97403-1222 email: `cwc@darkwing.uoregon.edu`

**Algebraic geometry and commutative algebra** to LAWRENCE EIN, Department of Mathematics, University of Illinois, 851 S. Morgan (M/C 249), Chicago, IL 60607-7045; email: `ein@uic.edu`

**Algebraic topology and cohomology of groups** to STEWART PRIDDY, Department of Mathematics, Northwestern University, 2033 Sheridan Road, Evanston, IL 60208-2730; email: `priddy@math.nwu.edu`

**Combinatorics and Lie theory** to SERGEY FOMIN, Department of Mathematics, University of Michigan, Ann Arbor, Michigan 48109-1109; email: `fomin@math.lsa.umich.edu`

**Complex analysis and complex geometry** to DUONG H. PHONG, Department of Mathematics, Columbia University, 2990 Broadway, New York, NY 10027-0029; email: `phong@math.columbia.edu`

***Differential geometry and global analysis** to LISA C. JEFFREY, Department of Mathematics, University of Toronto, 100 St. George St., Toronto, ON Canada M5S 3G3; email: `jeffrey@math.toronto.edu`

***Dynamical systems and ergodic theory** to ROBERT F. WILLIAMS, Department of Mathematics, University of Texas, Austin, Texas 78712-1082; email: `bob@math.utexas.edu`

**Functional analysis and operator algebras** to BRUCE E. BLACKADAR, Department of Mathematics, University of Nevada, Reno, NV 89557; email: `bruceb@math.unr.edu`

**Geometric topology, knot theory and hyperbolic geometry** to ABIGAIL A. THOMPSON, Department of Mathematics, University of California, Davis, Davis, CA 95616-5224; email: `thompson@math.ucdavis.edu`

**Harmonic analysis, representation theory, and Lie theory** to ROBERT J. STANTON, Department of Mathematics, The Ohio State University, 231 West 18th Avenue, Columbus, OH 43210-1174; email: `stanton@math.ohio-state.edu`

***Logic** to THEODORE SLAMAN, Department of Mathematics, University of California, Berkeley, CA 94720-3840; email: `slaman@math.berkeley.edu`

**Number theory** to MICHAEL J. LARSEN, Department of Mathematics, Indiana University, Bloomington, IN 47405; email: `larsen@math.indiana.edu`

***Ordinary differential equations, partial differential equations, and applied mathematics** to PETER W. BATES, Department of Mathematics, Brigham Young University, 292 TMCB, Provo, UT 84602-1001; email: `peter@math.byu.edu`

***Partial differential equations and applied mathematics** to BARBARA LEE KEYFITZ, Department of Mathematics, University of Houston, 4800 Calhoun Road, Houston, TX 77204-3476; email: `keyfitz@uh.edu`

***Probability and statistics** to KRZYSZTOF BURDZY, Department of Mathematics, University of Washington, Box 354350, Seattle, Washington 98195-4350; email: `burdzy@math.washington.edu`

***Real and harmonic analysis and geometric partial differential equations** to WILLIAM BECKNER, Department of Mathematics, University of Texas, Austin, TX 78712-1082; email: `beckner@math.utexas.edu`

**All other communications to the editors** should be addressed to the Managing Editor, WILLIAM BECKNER, Department of Mathematics, University of Texas, Austin, TX 78712-1082; email: `beckner@math.utexas.edu`.

## Selected Titles in This Series

*(Continued from the front of this publication)*

702 **Ilijas Farah,** Analytic guotients: Theory of liftings for quotients over analytic ideals on the integers, 2000

701 **Paul Selick and Jie Wu,** On natural coalgebra decompositions of tensor algebras and loop suspensions, 2000

700 **Vicente Cortés,** A new construction of homogeneous quaternionic manifolds and related geometric structures, 2000

699 **Alexander Fel′shtyn,** Dynamical zeta functions, Nielsen theory and Reidemeister torsion, 2000

698 **Andrew R. Kustin,** Complexes associated to two vectors and a rectangular matrix, 2000

697 **Deguang Han and David R. Larson,** Frames, bases and group representations, 2000

696 **Donald J. Estep, Mats G. Larson, and Roy D. Williams,** Estimating the error of numerical solutions of systems of reaction-diffusion equations, 2000

695 **Vitaly Bergelson and Randall McCutcheon,** An ergodic IP polynomial Szemerédi theorem, 2000

694 **Alberto Bressan, Graziano Crasta, and Benedetto Piccoli,** Well-posedness of the Cauchy problem for $n \times n$ systems of conservation laws, 2000

693 **Doug Pickrell,** Invariant measures for unitary groups associated to Kac-Moody Lie algebras, 2000

692 **Mara D. Neusel,** Inverse invariant theory and Steenrod operations, 2000

691 **Bruce Hughes and Stratos Prassidis,** Control and relaxation over the circle, 2000

690 **Robert Rumely, Chi Fong Lau, and Robert Varley,** Existence of the sectional capacity, 2000

689 **M. A. Dickmann and F. Miraglia,** Special groups: Boolean-theoretic methods in the theory of quadratic forms, 2000

688 **Piotr Hajłasz and Pekka Koskela,** Sobolev met Poincaré, 2000

687 **Guy David and Stephen Semmes,** Uniform rectifiability and quasiminimizing sets of arbitrary codimension, 2000

686 **L. Gaunce Lewis, Jr.,** Splitting theorems for certain equivariant spectra, 2000

685 **Jean-Luc Joly, Guy Metivier, and Jeffrey Rauch,** Caustics for dissipative semilinear oscillations, 2000

684 **Harvey I. Blau, Bangteng Xu, Z. Arad, E. Fisman, V. Miloslavsky, and M. Muzychuk,** Homogeneous integral table algebras of degree three: A trilogy, 2000

683 **Serge Bouc,** Non-additive exact functors and tensor induction for Mackey functors, 2000

682 **Martin Majewski,** ational homotopical models and uniqueness, 2000

681 **David P. Blecher, Paul S. Muhly, and Vern I. Paulsen,** Categories of operator modules (Morita equivalence and projective modules, 2000

680 **Joachim Zacharias,** Continuous tensor products and Arveson's spectral $C^*$-algebras, 2000

679 **Y. A. Abramovich and A. K. Kitover,** Inverses of disjointness preserving operators, 2000

678 **Wilhelm Stannat,** The theory of generalized Dirichlet forms and its applications in analysis and stochastics, 1999

677 **Volodymyr V. Lyubashenko,** Squared Hopf algebras, 1999

676 **S. Strelitz,** Asymptotics for solutions of linear differential equations having turning points with applications, 1999